PERGAMON I ~~University of South Wales~~ **Y**
of Science, Techr **Studi**

The 1000-volume o
industrial training an the enjoyment
Publisher: Robert Maxwell, M.C.

2045842

D1494681

ONE WEEK

Renew Books on PHC

Books are to be returned

THE PERGAMON TEXTBOOK
INSPECTION COPY SERVICE

Other Titles of Interest

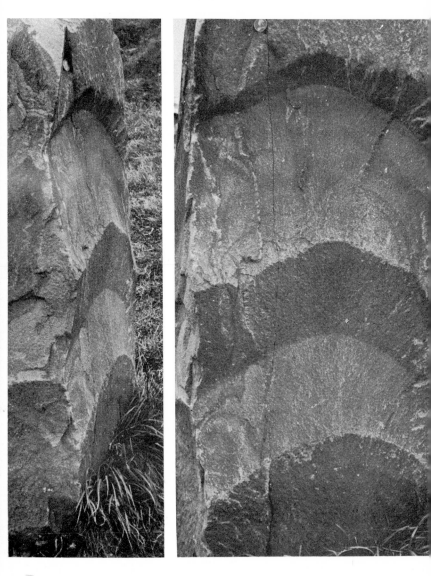

Frontispiece. Profile and front view of joint surface, developed in coal measure sandstone, showing large scale rib structure.

FAULT AND JOINT DEVELOPMENT

DEVELOPMENT

in Brittle and Semi-brittle Rock

BY

NEVILLE J. PRICE, Ph.D., B.Sc., F.G.S.

Department of Geology,
Imperial College of Science and Technology, London

PERGAMON PRESS

OXFORD · NEW YORK · TORONTO · SYDNEY · PARIS · FRANKFURT

U.K.	Pergamon Press Ltd., Headington Hill Hall, Oxford OX3 0BW, England
U.S.A.	Pergamon Press Inc., Maxwell House, Fairview Park, Elmsford, New York 10523, U.S.A.
CANADA	Pergamon Press Canada Ltd., Suite 104, 150 Consumers Rd., Willowdale, Ontario M2J 1P9, Canada
AUSTRALIA	Pergamon Press (Aust.) Pty. Ltd., P.O. Box 544, Potts Point, N.S.W. 2011, Australia
FRANCE	Pergamon Press SARL, 24 rue des Ecoles, 75240 Paris, Cedex 05, France
FEDERAL REPUBLIC OF GERMANY	Pergamon Press GmbH, 6242 Kronberg-Taunus, Hammerweg 6, Federal Republic of Germany

First edition 1966
Reprinted 1975, 1981
Library of Congress Catalog Card No. 65-27361

Printed in Great Britain by A. Wheaton & Co. Ltd., Exeter
0-08-011275-7 (Hardcover)
0-08-011274-9 (Flexicover)

Dedicated to Gilbert Wilson

CONTENTS

FOREWORD

THIS book, which deals with theories of fault and joint development in rock when they behave as brittle or semi-brittle material, is primarily intended for senior undergraduates and postgraduates in geology interested in the interpretation of geological structures; however, it may also be of interest to some mining and civil engineers.

The first chapter deals with some of the concepts and criteria of brittle failure, and an attempt is made to define limits of temperature and pressure below which rocks may behave in a brittle or semi-brittle manner. The second and third chapters deal with the application of these concepts of brittle failure and elastic theory to the problems of faulting and jointing respectively.

Since this book is primarily intended for students with some knowledge of structural geology, the mode of occurrence of faults in the field is only briefly mentioned. However, in view of the confusion which sometimes arises in the technical literature regarding joints, I have felt it necessary to deal with the idealized field relationships of these to other tectonic structures in greater detail.

It is emphasized that since this book deals with theoretical aspects of structural geology it is concerned with generalizations and approximations. It must be left to the reader to decide how closely these theories approximate to any particular field data.

I am indebted to various authors, editors and publishers for their permission to reproduce certain diagrams and plates and to summarize arguments. The publishing houses, journals and societies who furnished such permission are *American Journal of Science, Colliery Engineering, Geological Magazine,* Geological Society of America, *International Journal of Rock Mechanics, Journal of*

SECOND FOREWORD AND
SUPPLEMENTARY REFERENCE LIST

Almost two decades have passed since I started to write this book. In that time there have been significant advances in some of the topics covered by it.

Although Hubbert and Rubey wrote their classical paper on the influence of fluid pressure on overthrusting in the late fifties, it took many years for the importance of this concept to be applied more widely and to be fully appreciated by geologists. An up-to-date treatment of the mechanical (and chemical) importance of fluids in rock deformation generally is given by Fyfe, Price and Thompson (1978). This reference also covers aspects of rock mechanics, hydraulic fracture and migration of fluids in the crust in non-tectonic environments. It mentions the overthrusting problem only in passing. Up-to-date treatments on various aspects of thrust emplacement are to be found in McClay and Price (1981), a volume which contains the proceedings of a 3-day conference on Thrust and Nappe Tectonics held in Imperial College in 1979.

Unfortunately, a comparable comprehensive study of secondary (second order) fractures associated with a primary (first order) fault has not yet been published. The mechanics of development of secondary fractures has been considered by Chinnery (1966), Lajtai (1968) and Price (1968).

The situation in the literature regarding "joints" is far from satisfactory. There is no widely agreed definition of what the word means. As a result "joint" is used to describe a wide range of fracture types. Indeed the confusion on this topic is nowadays so widespread that I try not to use the word, and regret that it is included in the title of this book. The development of fractures (many of which would be termed joints by many geologists) in sediments which have not experienced tectonic deformation is given by Price (1974), and additional information regarding the topic of fracture ("joint") frequency in sediments is presented by Ladeira and Price (1981).

The few references which are cited above and listed below contain a much larger number of references which will lead the interested reader into the more recent literature.

References

Chinnery, M. A. 1966. Secondary faulting. *Can. J. Earth Sci.* **3**, No. 2, pp. 163-190.

Fyfe, W. S., Price, N. J. and Thompson, A. B. 1978. *Fluids in the Earth's Crust.* Elsevier, Amsterdam.

Ladeira, F. L. and Price, N. J. 1981. Relationship between fracture spacing and bed thickness. *Structural Geology* (in press).

Lajtai, E. Z. 1968. Brittle fracture in direct shear and the development of second-order faults and tension gashes. In *Proc. Conf. on Research in Tectonics,* pp. 96-111. Eds. Baer, A. J. and Norris, D. K. G.S.C. Paper 68-52.

McClay, K. and Price, N. J. 1981. *Thrust and Nappe Tectonics.* Geol. Soc. of London Spec. Publ. No. 9.

Price, N. J. 1968. A dynamic mechanism for the development of second-order faults and related structures. In *Proc. Conf. on Research in Tectonics,* pp. 49-71. Eds. Baer, A. J. and Norris, D. K. G.S.C. Paper 68-52.

Price, N. J. 1974. This development of stress systems and fracture patterns in undeformed sediments. *Advances in Rock Mechanics* pp. 487-496. Proceedings of the 3rd ISRM, Denver.

CHAPTER 1

BRITTLE FRACTURE

INTRODUCTION

Shear fractures, which are the result of differential movement of rock masses along a plane, are commonly observed in the field. Such structures are generally of tectonic origin and may range in size from faults with an extent of many tens of miles to small scale structures observable in hand specimen or under the microscope.

Occasionally, two intersecting shear planes or sets of planes, that is, *conjugate* or *complementary shears,* are encountered. When these are developed in competent rock they intersect as shown in Fig. 1, and the movement along the planes is such that the acute wedges between the shears have moved towards the line of intersection of the shear planes. There is abundant field evidence to show that the direction of the greatest compression at the time when these structures developed, intersected the acute angle formed by these planes.

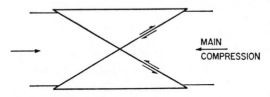

MAIN
COMPRESSION

Fig. 1. Conjugate shear fractures in a competent unit.

The type and orientation of shear planes represented in Fig. 1 are comparable with fractures generated in rock specimens during short term compression experiments conducted in the laboratory. It is usual to invoke one or other of the criteria of brittle failure when interpreting the orientation and development of such experimentally produced shears. Many geologists have assumed that these same criteria may also be used to interpret the development and orientation of faults and joints in the field.

Now it has been demonstrated in short duration tests in the laboratory, when specimens are at room temperature and under little or no lateral constraint, that competent rocks behave as brittle materials. However, in tests conducted at high confining pressure and high temperature, or when the load is applied for long periods, rock behaves in a manner which is other than brittle.

In the earth's crust, rock will, in fact, be under considerable constraint and at elevated temperatures: moreover, tectonic processes causing deformation will act over very long periods. Consequently, it is pertinent to enquire whether brittle failure criteria can legitimately be applied in environmental conditions of temperature, pressure and time comparable with those obtaining in the earth's crust; while, if it can be demonstrated that there exists in the crust a zone in which competent rocks approximate to brittle materials, it is of interest to attempt to define its limits in terms of pressure and temperature and therefore in terms of depth, for various types of competent rock. This chapter is given over to such an inquiry. However, since the book as a whole is concerned primarily with the behaviour of brittle materials, it is apposite at this point to deal with some of the simpler concepts of stress, strain and elasticity.

STRESS

If one considers a simple prism of unit cross-section area a subjected to a force F, as shown in Fig. 2, the stress σ_z acting on the end surface $ABCD$ is given by

$$\sigma_z = F/a. \tag{1}$$

Since the force F has no component acting parallel to the $ABCD$ surface it exerts no traction on this end surface, or in other words, the shear stress on this surface is zero. By definition, any stress acting perpendicular to a surface along which the shear stress is zero is a principal stress, i.e. σ_z is a principal stress. The suffix z indicates that the stress acts in the z-direction which, by convention,

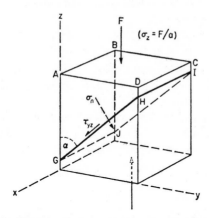

Fig. 2. Normal and shear stresses acting on external and internal surfaces of a small unit cube subjected to a compressive force F.

is usually taken as vertical. Other principal stresses may be orientated parallel to the x- and y-axes and would be designated σ_x and σ_y.

If the relative intensities of the principal stresses are known, they may be termed the maximum (or greatest), intermediate and minimum (or least) principal stresses (i.e. σ_1, σ_2 and σ_3 respectively).

If one considers the action of the force F on a surface $GHIJ$ inclined at an angle a, as indicated in Fig. 2, the component of F acting normal to $GHIJ$ is given by

$$F_n = F \cdot \sin a.$$

However, it will be seen that the area a' of $GHIJ$ is greater than the area a of $ABCD$, and that

$$a' = a/\sin a.$$

Therefore, the normal stress σ_n acting on the inclined surface is

$$F_n/a' = (F/a) \sin^2 a$$

so that

$$\sigma_n = \sigma_z \cdot \sin^2 a. \tag{2}$$

Similarly, the component of F tangential to the inclined plane is given by

$$F_T = F \cdot \cos a.$$

As before, $a' = a/\sin a$; consequently, the shear stress (τ) acting along the plane equals

$$F_T/a' = (F/a)\cos a \cdot \sin a \qquad (3)$$

i.e. $\qquad\qquad \tau = \sigma_z \cdot \cos a \cdot \sin a.$

Shear stresses as well as other stresses may be related to co-ordinate axes. Thus, the shear stress indicated in Fig. 2 acts in the yz plane and is designated τ_{yz}.

It may readily be inferred from eqns. (2) and (3) that the stresses acting normal and tangential to a plane in a biaxial stress field are represented by

$$\sigma_n = \sigma_1 \sin^2 a + \sigma_3 \cos^2 a \qquad (4)$$

and

$$\tau = (\sigma_1 - \sigma_3) \sin a \cdot \cos a. \qquad (5)$$

These equations may also be written in terms of the double angle ($2a$):

$$\sigma_n = \frac{\sigma_1 + \sigma_3}{2} - \frac{\sigma_1 - \sigma_3}{2} \cdot \cos 2a \qquad (6)$$

and

$$\tau = \frac{\sigma_1 - \sigma_3}{2} \cdot \sin 2a. \qquad (7)$$

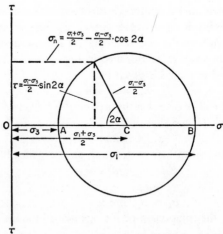

Fig. 3. Representation of stresses on any plane in a two-dimensional stress system by means of Mohr's stress circle.

These equations may be represented graphically by means of a Mohr's stress circle in which the normal stress (σ_n) and shear stress (τ) are chosen as co-ordinate axes (see Fig. 3). Here the greatest or maximum principal stress σ_1 and the least or minimum principal stress σ_3 are represented by OB and OA respectively on the σ_n axis. The quantity $(\sigma_1 + \sigma_3)/2$ represents the mid-point C on the abscissa between A and B, while $(\sigma_1 - \sigma_3)/2$ represents half the distance between A and B. If a circle is drawn with centre C and radius $(\sigma_1 - \sigma_3)/2$, then, for any specific values of σ_1 and σ_3, this circle represents the conditions of eqns. (6) and (7) where $2a$ is measured as indicated in Fig. 3. This construction, as we shall see, is used in representing the values of shear and normal stresses at failure.

It should be noted that in this book compressive stresses are regarded as positive quantities. This is the convention adopted by many geologists because the stresses in the earth's crust are usually compressive and it is more convenient to deal with positive quantities. However, it is as well to remember that the opposite sign convention is normally adopted by physicists and engineers.

ELASTICITY

Deformation prior to brittle failure is assumed to be perfectly elastic, i.e. when a load is applied to a body, the resulting deformation is said to be elastic if, when the load is removed, the deformation completely and instantly disappears.

If a perfectly elastic cube of side length l is subjected to a uniaxial compressive stress σ_z (see Fig. 4a, where the cube is presented in section), the cube shortens in the direction of the applied stress by an amount dl_z. The strain ε_z in the vertical direction due to the stress σ_z is then defined as

$$\varepsilon_z = dl_z/l. \tag{8}$$

It is usual, and mathematically convenient, to assume that the relationship between stress and strain is linear (Hooke's law) so that strain is related to stress by a constant E known as Young's modulus, where

$$E = \sigma_z/\varepsilon_z. \tag{9}$$

For an ideal, homogeneous and isotropic *Hookean solid E* possesses a constant value in tension, compression and in all directions within the solid.

In addition to the strain in the z-direction, the applied stress σ_z causes a lateral expansion $(dl_x = dl_y)$ in the x- and y-directions respectively. The ratio of the deformations dl_x/dl_z is termed Poisson's ratio v and the reciprocal dl_z/dl_x is known as Poisson's number (m).

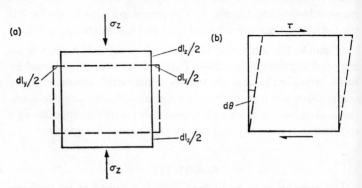

FIG. 4. (a) Vertical and horizontal strains in a unit cube (shown in section) induced by the stress σ_z. (b) Shear strain $(d\theta)$ in a unit cube induced by shear stress τ.

If a unit cube is submitted to triaxial compression, where σ_x, σ_y and σ_z are principal stresses, the shortening in the z-direction due to σ_z is given by

$$\varepsilon_z = \sigma_z/E.$$

However, the stresses σ_y and σ_x both produce an elongation in the z-direction, so that

$$- \varepsilon_z = \sigma_y/m \cdot E$$

and

$$- \varepsilon_z = \sigma_x/m \cdot E.$$

Hence, providing the strains are small, the total strain in the z-direction due to the triaxial compression is given by

$$\varepsilon_z = \varepsilon_{z_1} - (\varepsilon_{z_2} + \varepsilon_{z_3}) \tag{10}$$

$$= \frac{1}{E}\left[\sigma_z - \frac{1}{m}(\sigma_x + \sigma_y)\right].$$

The strains in the x- and y-directions can be obtained in a similar manner.

If an elastic cube is deformed by a shear stress, as indicated in Fig. 4b, the cube is deformed and becomes a rhomb. The small angle dθ, which the side of the rhomb AC makes with the side of the cube AB, is defined as the shear strain. The ratio

$$\frac{\text{shear stress}}{\text{shear strain}} = \frac{\tau}{\text{d}\theta} = G \tag{11}$$

where G is the modulus of rigidity.

Another elastic constant, known as the bulk modulus or compressibility, is a measure of the ratio of the change in volume of an elastic solid to the change in hydrostatic pressure, i.e.

$$K = \text{d}V/\text{d}P \tag{12}$$

where V is the volume and P the hydrostatic stress. These constants E, m, G and K are not independent, for if any two are known it is possible to calculate the remaining two. Young's modulus and Poisson's number are the constants which can most readily be determined experimentally. The expressions relating E and m to G and K are

$$G = \frac{m \cdot E}{2(m + 1)} \tag{13}$$

and

$$K = \frac{3(m - 2)}{m \cdot E}. \tag{14}$$

The derivation of these relationships is quite simple and will be found in any standard text dealing with elasticity (see Jaeger, Timoshenko, etc.).

Rocks only approximate to this concept of an ideal Hookean solid. For example, the stress–strain relationships are not invariably linear; consequently Young's modulus is not a simple constant, but is related to the applied level of stress. Moreover, since rocks are frequently non-isotropic, Young's modulus can vary depending upon the orientation of the fabric, or plane of anistropy with respect to the applied stress. For example, in sedimentary rocks, Young's modulus parallel to the bedding is higher than that perpendicular to the bedding (Price, 1958). In addition, Young's modulus is dependent upon the sign of stress and is greater in compression than it is in tension. Other deviations from the behaviour of an ideal Hookean solid will become evident later.

Some typical values for Young's modulus and Poisson's number for a number of representative rock types are tabulated below.

TABLE 1

Rock Type	Young's Modulus	Poisson's Number
Granite	$4\text{–}10 \times 16^6 \text{lb/in}^2$	$3-6$
Basalt	$8\text{–}12 \times 10^6 \text{lb/in}^2$	$3-6$
Sandstone (low porosity)	$6\text{–}11 \times 10^6 \text{lb/in}^2$	$4-10$
Dolomite	$7\text{–}10 \times 10^6 \text{lb/in}^2$	$3-6$

STRENGTH OF BRITTLE MATERIAL

The concept of strength as applied to a brittle material is, in the abstract, an extremely simple one. The behaviour pattern of such a material up to the moment of failure is assumed to be perfectly elastic. Consequently, the compressive and tensile strengths are respectively defined as the maximum applied compressive and tensile stresses that the body can withstand before failure occurs. In this context, failure means rupture of the specimen which often disintegrates with explosive violence.

In practice this concept of strength proves to be far from simple. In fact, even in the case of the so-called uniaxial compression test, experimental data which purport to represent the strength of a brittle test specimen depend upon so many factors that, unless the conditions of test and test material are known, the data are virtually useless and may be very misleading.

Test conditions which influence the "measured strength" include the rate of loading, the shape and relative dimensions of the specimens, the "end conditions" of the test piece and (when the test specimen is rock), the amount of moisture present in the pore spaces.

The influence of the loading rate is well established; the greater the loading rate and the induced rate of strain, the higher is the strength. However, for the loading rates normally employed in the laboratory the effect is not great (Obert *et al.*). Nevertheless, in short duration tests it is usual to standardize the rate of loading (usually about 100 lb/sec).

Strength is also related to specimen shape. Thus, for a given rock type and test technique, a unit cube has a greater strength than a unit cylinder (i.e. a cylinder with a length/diameter ratio of unity). Similarly, a unit cylinder is stronger than a cylindrical specimen with a length/diameter ratio of greater than unity.

In addition, the conditions at the end faces of the specimen have a great influence on strength, for they should be parallel to each other and normal to the axis of the specimen. The flatness and finish of the end faces are also very important. The writer has found that ridges and hollows on the end surfaces with amplitudes of only a few thousandths of an inch, cause the specimen to fail at loads much lower than those which can be supported by specimens with accurately ground end faces. Ridges and hollows cannot be accurately reproduced from specimen to specimen and are, therefore, one of the factors which will give rise to a scatter of results.

In an attempt to reduce this spread of results, it has sometimes been the practice to place a thin pad of some soft material, such as paper, cardboard, plywood or a ductile material, such as lead or copper, between the rock specimen and the metal platten of the press. All such materials produce tensile stresses at the interfaces which reduce, by a significant and often considerable amount, the strength relative to that obtained from tests in which a specimen, with carefully prepared end faces, is placed in direct contact with the metal plattens.

However, a rock–steel contact also influences the strength, for during a compression test the rock tends to expand laterally to a greater degree than the steel platten. Thus, shearing stresses are set up at the interfaces which tend to restrict expansion of the rock. It is possible to reduce the influence of these shearing stresses by means of a lubricant (Dreyer), such as graphite grease, but this material will penetrate into rock, especially the more porous rocks, and may thereby influence the measured strength.

Moisture in the pore spaces of the rock is also of great importance. In an air-dry rock specimen water is retained in the pore spaces by adsorption on all internal surfaces or in the form of capillary water. The total amount of water which is retained in this way is determined by the shape and size of the pores which, in turn, are related to the dimensions, geometry and packing of the grains. In a fine-grained rock, water may fill a substantial proportion of the pore space; but in a coarse-grained rock it will occupy only a small percentage of the pore volume.

The moisture content of various sandstones with differing porosities and the influence of this moisture on the strength of the rock is represented in Table 2.

TABLE 2

Rock type	Porosity (% vol.)	Air-dry pore-water content (% pore vol.)	Relative strengths		
			Completely dry (% σ_D)	Air-dry (% σ_D)	Saturated (% σ_D)
Pennant Sandstone	2·5	42±3	100	51	45
Markham Sandstone	6·0	22±5	100	57	—
Parkgate Rock	10·0	9±1·5	100	68	45
Darley Dale Sandstone	19·5	3±1·0	100	80	45

The oven-dry strength relates to specimens dried at 105°C until no further loss in weight of the specimens was observed. The air-dry and saturated strengths are expressed as percentages of the oven-dry strength for purposes of comparison.

It will be apparent from these remarks that the "simple compression test" is far from simple. In fact, it is probable that no test yet devised provides the strength of rock in uniform uniaxial compression. However, because the test is so easy to conduct, it is a useful empirical measure of rock strength.

INFLUENCE OF PETROLOGY ON UNIAXIAL STRENGTH

The data presented in Table 2 indicate that even though a number of rock types of different porosities are stored in the same conditions and crushed by means of the same technique, a direct comparison of the strength data may be most misleading, since, due to their varying pore-water contents, the rocks have, in fact, been tested in vastly different conditions. Clearly, when comparing strength data of various rocks it is essential to compare data of those rock types with similar pore-water contents or by using the moisture-free strength of the rocks.

In addition to the data presented in Table 2, the writer established that for Pennant Sandstone the mean strengths when water occupies 38 and 62 per cent of the pore spaces are 21,000 lb/in^2 and 19,000 lb/in^2 respectively. That is, within these limits, variations of pore-water content result in strength variations of ± 5 per cent about the mean value of 20,000 lb/in^2.

When studying the strength of a suite of coal measure rocks, the writer found that a large proportion of the rock types had porosities of 1–3·5 per cent and that the air-dry moisture content of the pore spaces was 38–62 per cent. If the average moisture content is taken as 50 per cent, then it can reasonably be assumed that the range of moisture contents of these rock types will give rise to variations in strength of 5 per cent. Thus, within these limits, the strength data obtained on these rock types can be used directly to demonstrate how the various petrological characteristics and other factors fundamentally determine the strength of these sedimentary rocks.

However, before doing this it is necessary to consider briefly the petrology of these rock types. The coal measure sandstones and siltstones tested were typical sub-greywackes (as defined by Pettijohn). They were composed of relatively poorly rounded and poorly sorted grains fixed in a very fine-grained matrix. The grains were dominantly quartz, but very small percentages of feldspar, mica, magnetite, pyrite and calcite, dolomite or sericite were sometimes present. The rocks can be divided into two series, depending upon whether the grains were fixed in a matrix comprised of clay mineral paste or in a cement dominantly composed of clay minerals plus calcite and/or other carbonate minerals.

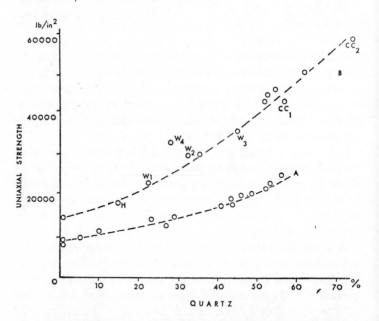

FIG. 5. Relationship between quartz content and uniaxial strength. Curve *A* refers to rock with clay mineral matrices whose strengths have been corrected to eliminate the effects of compaction. Curve *B* represents rock types with carbonate matrices.

From an inspection of the strength data, it was obvious that sandstones are, in general, stronger than mudstones, so that strength is likely to be influenced by the quartz content. In addition, it was noted that sandstones with a clay mineral matrix were, in general, weaker than those cemented sandstones with a similar quartz content, so that strength is obviously influenced by the type of matrix or cementing material.

The relationship between quartz content and uniaxial air-dry strength of those low-porosity (<1.5 per cent) coal measure rocks which possess a matrix containing carbonate minerals is shown in Fig. 5, where a definite, slightly curved relationship between strength and the quartz content can be seen.

The relationship between strength and the quartz content of those rocks with a matrix of clay minerals and with porosities less than 3.5 per cent is shown in Fig. 6.

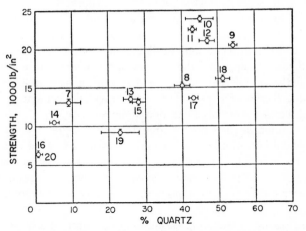

Fig. 6. Relationship between uniaxial strength and quartz content for rocks with clay mineral matrices.

In general, the higher the quartz content the stronger the rock, but the relationship is far less well defined than that exhibited by the calcareous series. It is suggested that the scatter can largely be attributed to another main variable factor, namely, compaction.

When sediments are deposited, individual grains, although in contact with neighbouring grains, are discrete and the sediments are without cohesion. As sediments become buried beneath later deposits they compact. For the majority of sediments this compaction results in interlocking of grains accompanied by fusion of the grains at their points of contact, so that the rock takes on cohesive strength. The greater the compaction, the greater is the fusion and, therefore, the greater the cohesion. Hence, the strength of many of these coal measure rocks will be related to the maximum compressive forces (due to depth of burial and possibly tectonic influences) and the temperatures to which they have been subjected.

The factors which brought about this compaction were also instrumental in devolatizing the coal seams. The empirical relationship between depth of cover of the seams (relative to an arbitrary datum) and their volatile content is well known. Published data (Jones, 1951) show that in the South Wales coalfield there is a linear relationship between volatile content and depth of cover between the limits of 12·5 – 40 per cent volatiles. If it is assumed that the degree of compaction of rocks exhibits a similar relationship in this and other coalfields, since the samples of the suite of rocks being discussed were collected underground adjacent to specific seams at specific collieries, it is possible to use the volatile content of the seam as a measure of the compaction of the rock.

The relationship between the strength and the degree of compaction obtained by plotting the strength of rock against the volatile content of the seam is given in Fig. 7. Clearly, for a given quartz content, the smaller the volatile content of the associated seam, i.e. the higher the degree of compaction, the greater is the strength of the rock.

Since we are interested in establishing the depth of any brittle-zone which may exist, it is of interest to digress at this point. All the rock types represented in Fig. 7 were obtained outside the Anthracite Zone, from areas where intense folding and other evidence of considerable lateral pressure is absent. It may therefore be assumed, that the influence of tectonic compression was small and that the degree of compaction and strength of these particular sediments are related to vertical and horizontal pressures due to gravitational loading and to temperature; factors which are, in this instance, related to, and increase in proportion to, the depth of cover. This assumption is, of course, open to debate, but the data given in Fig. 7 certainly lend support.

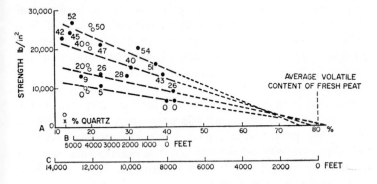

Fig. 7. Inter relationship between uniaxial strength, quartz content, compaction and depth of cover. Scale *A* indicates the degree of compaction as represented by the volatile content of the associated seam. Scale *B* gives the thickness of sediments separating the South Wales coal measure rocks from an arbitrary datum represented by the 40 per cent volatile level. Scale *C* shows the estimated maximum depth of cover for the various coal measure rocks.

If the strength–compaction relationship is linear, then the "quartz contours", when extrapolated, should intercept the abscissa at a single point. This point would represent unconsolidated sediments at the surface whose uniaxial strength is negligibly small (cf. soil mechanics tests).

It will be seen that, in fact, the quartz contours intercept the abscissa between 75–80 per cent volatile. In view of the scatter of experimental data and the length of the extrapolation, this is a reasonable approximation to intercepting "at a point". It is interesting to note that the average volatile content of fresh peat, i.e. peat which occurs at the surface and is uncompacted, is also 80 per cent (see Jones, 1951).

The rate of increase in volatile content of South Wales coal with increased depth of cover, below an arbitrary datum of 40 per cent volatiles, has been given by Jones and this is indicated by scale B in Fig. 7. Since the extrapolation of the quartz contours and the data of the average volatile content of fresh peat indicate that the original surface is probably represented by the 80 per cent volatile content mark, an estimate, based on strength and composition, can be made of the maximum depth of burial of these rocks (see scale C, Fig. 7). It must be pointed out that the data presented are such that the estimated maximum depth of burial may be subject to errors of $\pm$ 2000 ft.

It has been noted that the porosity of these rocks is less than 3·5 per cent. Now a depth–porosity relationship obtained by Arthy, indicates that shales attain such values of porosity at depths of 6000-7000 ft, or greater. Thus, there is general agreement between this depth–porosity relationship and the volatile–depth–strength relationship given in Fig. 7. It will be seen from Table 3 that these estimates are also in reasonable agreement with estimates based on various techniques made by Trotter, Jones (1951), Wellman, and Suggate.

TABLE 3

Author	Rank range	Depth (ft)
Jones	Peat—40 % vol.	8500
Wellman	Peat—Low rank bit.	8000
Trotter	Peat—Low rank bit.	6000-7000
Suggate	Peat—40 % vol.	10,000
Price	Peat—40 % vol.	8000
Price	Peat—10 % vol.	14,000
Suggate	Peat—10 % vol.	16,000
Jones	Peat— 5 % vol.	17,000
Suggate	Peat— 5 % vol.	19,000
Wellman	Peat— 4 % vol.	18,000

To continue with the study of the influence of petrology upon strength. The equivalent uniaxial strength of any of the rocks represented in Fig. 7, which any specific rock type would possess for any arbitrary degree of compaction, between the limits of 10 and 40 per cent volatiles, can be obtained by projecting the data parallel to the quartz contours. This enables the effect of compaction to be eliminated and permits the relationship between quartz content and uniaxial strength to be demonstrated. The relationship for a degree of compaction represented by the 25 per cent volatile content is represented by curve A in Fig. 5.

It is suggested that the straightforward relationship between quartz content and strength of the calcareous series, which is not masked by any compaction effect, can probably be related to the fact that once the cement has been deposited and has filled up almost all void spaces, the rock is, in effect, almost perfectly compacted.

It may be inferred from these relationships that the rocks with a carbonate matrix may possibly behave as a brittle material from the time that they have been completely cemented. However, the arenaceous rocks with a clay mineral matrix cannot be considered as brittle materials while they are undergoing compaction and "volume flow". But after their compaction, i.e. during a following phase of uplift or during any subsequent phase of subsidence, they may possibly behave as brittle materials.

INFLUENCE OF ENVIRONMENTAL FACTORS
ON BRITTLE FAILURE

It has been noted earlier that the behaviour of rock is known to be influenced by a number of factors which include confining pressure, temperature, presence or absence of pore-fluids and by the rate of strain induced in the specimen. It is of interest, therefore, to consider the influence of these various environmental factors on the behaviour of rock in laboratory experiments in order to determine, as far as possible, the conditions when rock passes out of the brittle state and behaves in a non-brittle manner.

Adams and his collaborators made the earliest study of rock subjected to constraint. In these experiments the specimens were constrained in heavy steel jackets. This technique had the obvious disadvantage that the confining pressure could not be measured, but had to be computed; moreover, the confining pressure varied continually throughout the tests.

The first experiments, in which relatively large hydraulic confining pressures of 2000 atm (i.e. approximately 29,500 lb/in²) were used, were those conducted by von Karmen.

Griggs and co-workers later extended the scope of this type of experiment and used confining pressures up to 13,000 atm (190,000 lb/in²) and also studied the influence on strength of time, temperature and various fluids. Similar studies have been reported in a series of papers by Handin and Hager and their associates and also by Serdengecti and Boozer (1961).

The type of apparatus used tends to become more sophisticated in the later experiments, with various modifications for measuring strain, heating the specimen and measuring internal pore-pressure. However, the "pressure capsule" is essentially as represented in Fig. 8.

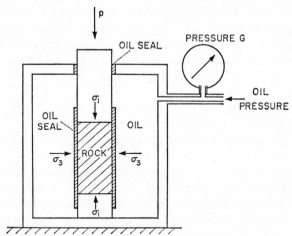

FIG. 8. Diagrammatic representation of a simple, triaxial compression "pressure capsule".

Tests are normally carried out on small, cylindrical specimens which are encased in a thin, flexible metal or rubber covering which separates the specimen from a fluid which provides the constraining pressure. The hydraulic confining pressure, which acts on the curved surface of the specimen, is maintained constant throughout the test and usually represents the least principal stress σ_3. The pressure acting parallel to the axis of the specimen builds up until the rock sample fails, when it represents the maximum principal stress σ_1 at failure.

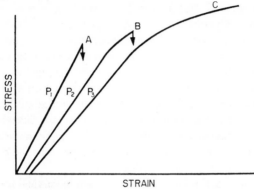

FIG. 9. Stress–strain curve of (A) brittle failure, (B) semi-brittle failure and (C) ductile failure. N.B.—these curves are not to scale.

It has been found by the various experimenters that at relatively low confining pressures P_1 the stress–strain relationship is typically brittle (see curve A in Fig. 9, where failure is represented by the vertical arrow). Curve B represents a transitional type of stress–strain relationship, being neither completely brittle nor completely ductile, which obtains at somewhat higher confining pressures P_2: such behaviour can be termed semi-brittle. Again, failure is indicated by an arrow. A typical ductile stress–strain relationship obtained at high confining pressures P_3 is represented by curve C.

For any one rock type there will be every gradation between curve A and curve C. In experimental work, detection of the onset of semi-brittle behaviour, as exemplified by curvature of the stress–strain curve, will largely depend upon the accuracy and sensitivity of the strain measuring device. Consequently, the practical limit set on brittle behaviour will, to some extent, be arbitrary and approximate.

The percentage elastic strain which a rock can exhibit before failure depends upon the rock type and the confining pressure. For example, Hobbs (1960) found that some coals exhibit up to 4 per cent elastic strain at a confining pressure of 4000 lb/in², and a differential pressure ($\sigma_1-\sigma_3$) of 20,000 lb/in², while Oil Creek Sandstone (Handin and Hager, 1957) exhibited an elastic strain of approximately 3 per cent at a confining pressure of 29,000 lb/in² and a differential pressure of approximately 160,000 lb/in².

Clearly, the precise limit can only be determined from a study of the stress–strain data. However, for convenience, one may follow Heard who, when studying the transition from brittle to ductile behaviour of Solenhofen Limestone, set the limit of brittle behaviour at 3 per cent. The transition from brittle to ductile behaviour, i.e. semi-brittle behaviour, was taken to occur for strains between 3 and 5 per cent. For amounts greater than 5 per cent the deformation was taken to be wholly ductile.

The pressure conditions at which various minerals and rock types cease to behave as a wholly brittle material in short term tests vary widely. For example, Griggs (1940) and Bridgman have conducted experiments on quartz at high confining pressures (but at temperatures of 20–25°C) and observed that it behaves as a brittle material even when subjected to confining pressures of 74,000 lb/in² and a differential pressure of 320,000 lb/in². Granite and basalt at 25°C are able to support a differential stress of 290,000 lb/in² and 250,000 lb/in² at a confining pressure of 74,000 lb/in² and still exhibit a strain of less than 5 per cent (Griggs et al.). A well-compacted sandstone (e.g. Oil Creek Sandstone) will remain semi-brittle at confining pressures greater than 29,000 lb/in². The compact, competent Blair Dolomite remains semi-brittle at similar confining pressures. However, Solenhofen Limestone (Heard) ceases to behave as a brittle material at confining pressures greater than 14,500 lb/in². In addition, many relatively porous sediments show more than 5 per cent permanent strain at quite low confining pressures (less than 7500 lb/in²). For these porous rock types, it seems likely that the permanent strain is due to compaction of the sediments at pressures which are comparable with, or even exceed, those pressures which obtained at their maximum depth of burial. In arenaceous rock this compaction takes place by cataclastic flow; that is, by displacement of one grain boundary relative to another, often accompanied by granulation and brittle fracture of the component grains. In limestone the compaction process may also be accompanied by twinning or translational gliding along preferred crystallographic planes.

Due to gravitational loading, the vertical stress in the earth's crust at any depth z measured from the surface is given by

$$\sigma_z = \rho_m \cdot g \cdot z \tag{15}$$

where ρ_m is the mean relative density of the rock mass and g is the gravitational acceleration. If it is assumed that ρ_m is reasonably constant for a sequence of rocks in the upper portion of the crust and has a value of 2·35, then the vertical stress increases by 1 lb/in² for every foot of cover; i.e. at a depth of 10,000 ft the vertical stress is approximately 10,000 lb/in². Hence, if confining pressure were the only environmental condition to affect brittle behaviour, granite would behave as a brittle material at depths as great as 75,000 ft. However, the experiments referred to above were conducted at room temperature at about 20°C, while at a depth of 75,000 ft it is reasonable to anticipate temperatures of the order of 750°C (assuming an average temperature gradient of 1°C per 100 ft increase in cover).

Such high temperatures would mitigate against brittle behaviour. For Griggs *et al.* (1958) found that igneous rocks such as granite, basalt and dunite behaved in a ductile manner at a confining pressure of 70,000 lb/in² and a temperature of 300°C.

However, a rock temperature of 300°C is likely to exist at depths of approximately 30,000 ft; so to simulate geological conditions, it is more realistic to associate this temperature with a confining pressure of 30,000 lb/in². At this confining pressure Handin and Hager found that the effect of heating was to reduce the strength of the rock. The magnitude of the effect, however, differed from one rock type to another. Oil Creek Sandstone, for example, is little affected; for the differential stress a specimen can sustain falls only from 160,000 lb/in^2 to 140,000 lb/in^2 when the temperature is increased from 24 to 300°C. For Luning Dolomite, however, the effect is more marked, for the strength decreased from 73,000 lb/in^2 to 33,000 lb/in^2 for a similar increase in temperature. However, it is important to note that for both rock types the mode of behaviour remained semi-brittle.

For some rock types the conditions marking the limit of brittle behaviour occur at lower temperatures and confining pressures. For example, although Solenhofen Limestone is essentially brittle at a temperature of 150°C and a confining pressure of 11,000 lb/in^2, at this same temperature and a confining pressure of 15,000 lb/in^2 the behaviour pattern is ductile.

The above data refer to jacketed specimens tested in the air-dry condition. In the earth's crust, however, pore and void spaces will, in general, be filled with some fluid which will, due to gravitational loading and possibly tectonic processes, be under pressure. This pore-water pressure has a mechanical effect which will be discussed in detail in a later chapter. It will then be shown that the important factor is the ratio of the pore-water pressure to confining pressure. As this ratio approaches unity, the level of confining stress at which failure is still brittle increases.

Such fluids may also be chemically active and thereby influence the behaviour pattern of rock material. For example, Griggs (1941) has demonstrated that a 10 per cent solution of sodium carbonate at 400°C reduces the strength of quartz by 80 per cent and that even in water, a high degree of preferred orientation can be induced by shearing quartz at this temperature under a confining pressure of 15,000 lb/in^2 in combination with a stress of 15,000 lb/in^2 acting normal to the shear plane.

Maxwell has inferred that quartz deformation in the presence of fluids is affected at lower temperatures. From his experiments on the compaction and cementation of sands, under pressures equivalent to a depth of overburden of 26,500 ft, he concludes that the plastic deformation of quartz begins to be important at about 235°C. Under these conditions plus relatively long periods of loading (some of his compaction experiments had a duration of 100 days) he suggests that redistribution of stresses within the rock mass is brought about by plastic deformation of the grain boundaries and not by granulation and brittle failure of the quartz grains.

It is interesting to note that Westbrook, using a Vickers indenter, found a "ductility maximum" in the vicinity of 200°C. Further, Serdengecti *et al.* (1962), when studying deformation of Navajo Sandstone, at a temperature of 200°C, reported that under a confining pressure of 20,000 lb/in², grains near a shear zone showed evidence of considerably more permanent strain than specimens sheared at a lower confining pressure.

Thus, from various data obtained in short duration experiments mentioned above, it may be taken that at a temperature exceeding 200–235°C and a confining pressure not much greater than 20,000 lb/in² quartz begins to deform in a plastic manner.

This conclusion appears to be at variance with the experimental data obtained by Griggs *et al.* (1958) when it was shown that solid "dry" cylinders of Brazilian Quartz behaved as a brittle material even when submitted to a confining and differential pressure of 74,000 lb/in² and 320,000 lb/in² respectively at a temperature of 800°C However, it must be emphasized that the conclusion reached in the preceding paragraph was based on experimental data obtained from specimens of granular and crystalline aggregates in contact with fluids. In such specimens, although the applied stress may be relatively modest, we shall see later that the micro stresses which act on a grain or a portion of a grain, can be an order of magnitude greater than the stresses applied to the solid quartz cylinders.

Some idea of the complexity and intensity of the local stress distribution in a material composed of irregularly shaped granules can be obtained from a photoelastic fringe-pattern set up in irregularly shaped, planar, pieces of Araldite, stuck together at their points of contact and subjected to uniaxial compression (see Plate 1). It is the action of these intense micro-stresses coupled with the presence of fluids which result in local plastic flow at relatively low temperatures and moderate values of applied stress.

Consequently, it may be concluded that those igneous, meta-morphic and sedimentary rocks in which quartz is an important mineral will begin to deform, in response to tectonic pressures, in a plastic manner at levels in the crust where these conditions of temperature and pressure obtain, i.e. at a depth greater than approximately 20,000 ft. Compacted, relatively non-porous limestones, it may be inferred, deform plastically at shallower depths, possibly under a cover of only 8–10,000 ft.

For arenaceous rocks, the field evidence appears to be in general agreement with these conclusions. Strain shadows in quartz grains from sediments from the Midland coalfields of England are relatively infrequent, while in sandstones in the anthracite area of South Wales (which, as we have seen, have probably had a maximum depth of cover of almost 20,000 ft) the quartz grains generally show intense strain shadows around grain boundaries and whole grains frequently exhibit undulose extinction.

Admittedly, the use of strain shadows as a measure of deformation is open to question (see Blatt and Christie); but the occurrence and production of undulose extinction in quartz is known to be associated with folds (Lowry) and faults (Christie and Raleigh). Moreover, Anvruddha and Maxwell suggest that such undulose extinction may also be produced by compaction alone. As a result, it is suggested that the strain shadows of the quartz grains in the sandstones from the anthracite area indicate that plastic deformation of the component grains, and hence of the rock as a whole, has begun at an estimated depth of approximately 20,000 ft.

RELATIONSHIPS BETWEEN PRINCIPAL STRESSES AT FAILURE

It has been found that, for a great many competent rock types, the relationship between principal stresses at failure (at moderate confining pressures) is adequately represented by the empirical equation

$$\sigma_1 = \sigma_0 + K\sigma_3, \tag{16}$$

where σ_0 is the uniaxial compressive strength and K is a constant which usually has a value of 2·0 to 15·0. Moderately competent argillaceous and arenaceous rock types generally have values of K between 2·0 and 5·0. The more competent arenaceous rocks and competent limestones generally have values ranging from 5·0 to 10·0, while igneous rocks commonly have values of K greater than 10·0.

The simple linear relationship between principal stresses for coal measure sandstone (Price, 1958) is shown in Fig. 10a. With many rock types, however, the relationship between principal

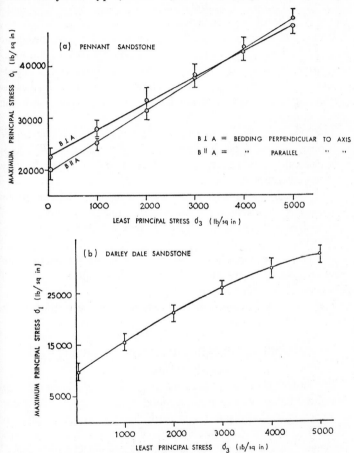

FIG. 10. Linear and curved relationships between principal stresses at failure.

stresses is clearly non-linear (see Fig. 10b). Balmer has suggested that such curved relationships can be expressed by the equation

$$\left. \begin{array}{l} \sigma_3 = A\sigma_1^{\,B} + C \\[2mm] \text{or} \\[2mm] \sigma_1 = \left[\dfrac{1}{A}(\sigma_3 - C) \right]^{\frac{1}{B}} \end{array} \right\} \tag{17}$$

where A, B and C are constants.

Clearly, if B equals unity, this expression reduces to the linear form given in eqn. (16), so that $K = 1/A$ and the uniaxial compressive strength $\sigma_0 = -C/A$. Now, it may be inferred that C represents the tensile strength of the rock type; hence, the uniaxial compressive strength can be expressed in terms of a constant times the tensile strength.

The values of these constants were established for a number of coal measure rock types by Price (1963). The value for B, which is of particular interest, ranged from 1·07 to 1·89 (or approximately between the limits of 1·0 and 2·0).

Various criteria have been proposed to explain the conditions which govern the failure of real materials under combinations of compressive stresses: a summary of which has been made by Robertson (1955) and Hobbs (1964a). Of these hypotheses and theories only the two whose predictions regarding the relationship between principal stresses at failure best satisfy the experimental data cited above will be considered here.

These are the Navier–Coulomb criterion and the Griffith crack theory of brittle failure.

NAVIER–COULOMB CRITERION OF BRITTLE FAILURE

This criterion of brittle shear failure is based upon the concept that shear failure will take place along a surface when the shear stress acting in that plane is sufficiently large to overcome the cohesive strength of the material plus the frictional resistance to movement. The frictional resistance to movement is said to equal the stress normal to the shear surface multiplied by the coefficient of the internal friction of the material; while the cohesive strength of the material is its inherent shear strength, when the stress normal to the shear surface is zero. Thus, the failure criterion, which appeals to the intuition, may be expressed as

$$\tau = S + \mu_i \sigma_n \qquad (18)$$

where τ is the shear stress acting along the shear surface, σ_n is the normal stress, S is the cohesive strength and μ_i represents the coefficient of internal friction. By analogy with ordinary sliding friction

$$\mu_i = \tan\varphi_i$$

where φ_i is the angle of internal friction. So that the shear criterion may be written

$$\tau = S + \sigma_n . \tan \varphi_i. \qquad (18a)$$

Considering the simple, two-dimensional case: the intermediate principal stress σ_2 acts parallel to the shear plane and at right angles to the direction of shear movement, it is therefore assumed to have no influence upon failure.

It has been indicated that stresses at failure σ_1 and σ_3 may be represented by Mohr's circles. The curve, which is tangent to such circles, represents the failure conditions for the material under test. Hence, when the envelope is linear, it has the equation of the Navier–Coulomb shear failure criterion (see Fig. 11).

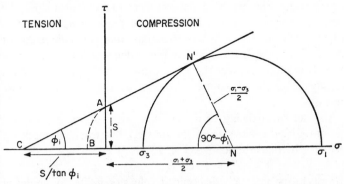

Fig. 11. Representation of the Navier–Coulomb criterion of brittle failure as a linear relationship between shear τ and normal σ_n stresses (after Hubbert and Rubey).

It can readily be shown that this criterion of failure gives rise to a linear relationship between principal stresses at failure. Thus, from Fig. 11 it follows that

$$\frac{\sigma_1 - \sigma_3}{2} = \left(\frac{\sigma_1 + \sigma_3}{2} + \frac{S}{\tan \varphi_i}\right) \sin \varphi_i$$

which simplifies to

$$\sigma_1 = \frac{2 \cdot S \cos \varphi_i}{1 - \sin \varphi_i} + \sigma_3 \left(\frac{1 + \sin \varphi_i}{1 - \sin \varphi_i}\right). \tag{19}$$

Thus, from eqns. (16) and (19), it follows that

$$K = (1 + \sin \varphi_i)/(1 - \sin \varphi_i). \tag{20}$$

Moreover, it can be shown that

$$\frac{\cos \varphi_i}{1 - \sin \varphi_i} \equiv \sqrt{\left(\frac{1 + \sin \varphi_i}{1 - \sin \varphi_i}\right)} \tag{21}$$

so that the uniaxial strength $\sigma_o = 2 \cdot S \cdot K^{\frac{1}{2}}$, i.e. the uniaxial compressive strength is related to the cohesive strength and the angle of internal friction.

The Navier–Coulomb criterion is in agreement with experimental data for a number of rock types, for which the relationship between principal stresses at failure is very nearly linear, i.e. B of eqn. (17) is approximately equal to unity. To account for a nonlinear relationship between principal stresses at failure, it is necessary to assume that the frictional resistance to movement along the shear plane is not related to the normal pressure σ_n in a linear fashion—that is, it is necessary to assume that either the angle of internal friction φ_i is not constant but is dependent upon pressure or, as is more likely, that the area of grains in the rock specimen, actually in frictional contact, increase as the normal pressure increases.

Another shortcoming of the criterion arises from the fact that it is not related to the sign of the stresses. The tensile strength T_p, predicted by this criterion (see Fig. 11), is given by

$$T_p = S \cdot \cot \varphi_i.$$

For most experimentally determined values of φ_i (i.e. for angles of φ_i less than 45°) it follows that the predicted tensile strength is often considerably larger than the cohesive strength. However, this is at variance with experimental data, for the measured tensile strength is always smaller than, and is frequently approximately half, the cohesive strength. Consequently, it is reasonable to assume that the envelope in the tensile quadrant of Fig. 11 is represented by the curved line *AB* and not by the straight line *AC*.

GRIFFITH CRITERION OF BRITTLE FAILURE

By considering the atomic forces necessary to cause failure, it is possible to show that the strength of an ideal brittle solid is given by

$$T \simeq E/10$$

where E is Young's modulus (see Freudenthal and Cottrell). For many competent rocks $E = 10^7$ lb/in². Consequently, the ideal strength should be of the order of 10^6 lb/in². Griffith suggested that the vast discrepancy between the theoretical and the observed values of the tensile strengths of materials was a result of the intense local stress concentrations which developed in the immediate vicinity of microscopic flaws. Although the mean stress throughout the body may be relatively low, these local stresses, according to the theory, attain values equal to the theoretical strength.

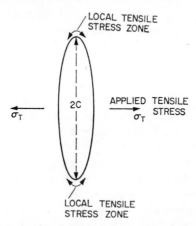

FIG. 12. "Griffith" elliptical flaw with its long axis normal to a tensile stress field σ_T.

Griffith, using a technique previously adopted by Inglis, calculated the stress concentration around an elliptical crack set in a two-dimensional thin plate, which is subjected to a tensile stress (see Fig. 12). The maximum tensile stress T_m associated with such an elliptical flaw is given by

$$T_m = 2\sigma_T \sqrt{\frac{c}{r_m}} \tag{22}$$

where σ_T is the applied tensile stress, $2c$ is the length of the major axis of the elliptical flaw and r_m is the radius of curvature of the flaw at the end of the major axis. Clearly, as r_m approaches zero, T_m tends to infinity. Hence, it is not possible to give T_m a definite value without making assumptions regarding the physical dimensions of the flaw.

When a crack propagates, part of the potential energy (excess elastic energy) is used in generating displacements around the flaw, which in turn give rise to vibrations (which are sometimes audible), and part in creating surface energy (analogous to the forces which result in surface tension in a liquid) at the freshly formed surface. A crack is in unstable equilibrium with the externally applied stresses if, for a small increase in length of the crack, the increase in surface energy and of the excess elastic energy is just equal to the work done by the external stresses. Griffith showed that this condition is attained when

$$\sigma_T = \sqrt{\frac{2 S_e . E}{\pi c}} \tag{23}$$

where S_e is the surface energy of the flaw. Since the equilibrium is unstable, the flaw will propagate as soon as σ_T is exceeded by an infinitesimal amount. Hence, this critical stress σ_T equals the macroscopic, or bulk, tensile strength T of the material.

Orowan has derived the Griffith equation for critical stresses (except for a minor difference in the numerical factor) by using atomic force concepts; thus putting the Griffith theory on a sound theoretical foundation.

By taking reasonable values for the physical constants in eqn. (23), Griffith obtained a value for the critical stress σ_T which was in good agreement with measured values for the tensile strength of large specimens of glass. He obtained even more striking evidence to support the theory by preparing "fibres" of glass which were almost completely free from flaws and surface imperfections. He found that the tensile strengths of such specimens were as much as 600,000 lb/in² which, of course, approximates reasonably closely to the theoretical "atomic" strength.

Griffith also dealt with the propagation of flaws in a sheet which was subjected to biaxial compression in the plane of the sheet. He assumed that a thin isotropic sheet contained a series of randomly orientated elliptical micro-flaws which were so spaced, one from the other, that the stress field associated with each flaw did not interfere with the stress field of the neighbouring flaw. Now it can be shown, that even when the applied stresses are compressive, some of the stresses associated with each flaw will be tensile. The mathematical argument leading to this conclusion is somewhat lengthy and, since it is given by Griffith, Jaeger, Odé, and others, will not

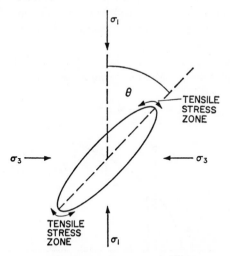

Fig. 13. "Griffith" elliptical flaw in a biaxial compressive stress field with its long axis making an angle θ with the axis of greatest principal stress σ_1.

be given here. It can be inferred from eqn. (22) that the stress magnification around the flaw is greatest where the radius of the curvature is smallest, namely, at the end of the major axis. Hence, the tensile stresses which develop around portions of the flaw will have the most profound influence when the tensile stress zone coincides with this zone of minimum radius of curvature. Griffith shows that this condition obtains when the angle θ, which the major axis of the ellipse makes with the axis of maximum principal stress (see Fig. 13), is given by

$$\cos 2\,\theta = -\ \tfrac{1}{2}\left(\frac{\sigma_1 - \sigma_3}{\sigma_1 + \sigma_3}\right). \tag{24}$$

Griffith also showed that, provided $\sigma_1 \neq \sigma_3$ and $3\sigma_1 + \sigma_3 < 0$, the tensile stresses around the flaw reach the critical stress resulting in the spreading of the crack and failure, when

$$(\sigma_1 - \sigma_3)^2 + 8T(\sigma_1 + \sigma_3) = 0. \tag{25}$$

It will be noted that when $\sigma_3 = 0$ the uniaxial compressive strength equals $8T$, so that, again, the uniaxial compressive strength is expressed in terms of the tensile strength. This predicted ratio of uniaxial compressive to tensile strengths is in good accord with experimental data.

This quadratic relationship between principal stresses at failure can be expressed as a Mohr's envelope (see Murrell, 1958) with the equation

$$\tau^2 + 4T\sigma_n - 4\,T^2 = 0. \tag{26}$$

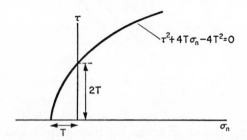

FIG. 14.   Relationship between shear and normal stresses at failure predicted by the Griffith criterion of brittle failure.

An envelope satisfying this equation is indicated in Fig. 14. It will be seen that the shape of the envelope in the tensile region is of the form inferred from experimental data. Also, the "cohesive strength" of the Navier–Coulomb shear failure criterion is equal to twice the tensile strength; a figure which is in good agreement with values estimated from experimental data. The shape of the envelope in the compressive region satisfies experimental data obtained by Hobbs (1964a).

Now it seems probable, that the micro-flaws in many rocks, in so far as they can be considered as approximate ellipses, will have great eccentricity. Consequently, in a compressive stress field these flaws will tend to close, or partially close, in an elastic manner. Terry has demonstrated that micro-cleats and other small scale fractures in coal behave in this manner.

McClintock and Walsh have used this concept to modify the simple Griffith theory. They considered partially closed cracks to have the form represented in Fig. 15. These flaws were assumed to close when the stress reached a certain value σ_c. Once the crack had closed, the effective normal stress across the crack is then given by

$$\sigma_e = (\sigma_n - \sigma_c).$$

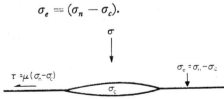

FIG. 15. Griffith flaw modified by McClintock and Walsh showing portion of flaw closed under external stress σ and supporting a shear stress τ

Since the crack surfaces are in contact, shear stresses may be generated along the closed portion of the crack and these are given by

$$\tau_e = \mu_s \sigma_e = \mu_s(\sigma_n - \sigma_c)$$

where μ_s is the coefficient of sliding friction.

McClintock and Walsh superimposed these stresses on the stress field around the flaw given by the solution derived by Griffith and obtained a relationship between principal stresses at failure of the form

$$\mu_s(\sigma_1 + \sigma_3 - 2\sigma_c) + (\sigma_1 - \sigma_3)(1 + \mu_s)^{\frac{1}{2}} = 4T(1 - \sigma_c/T)^{\frac{1}{2}}. \quad (27)$$

Now for many low porosity rock types, the pore spaces between grains will be extremely long, narrow spaces which are likely to close under very low compressive stresses, so that one may make the approximation that $\sigma_c = 0$. For such rock types, the relationship between principal stresses at failure given in eqn. (27) reduces to

$$\mu_s(\sigma_1 + \sigma_3) + (\sigma_1 - \sigma_3)(1 + \mu_s)^{\frac{1}{2}} = 4T. \quad (28)$$

This equation represents a linear relationship between principal stresses at failure.

Brace (1960) has shown that this equation has a Mohr's envelope for compressive stresses of the form

$$\tau = 2T + \mu_s \cdot \sigma_n. \quad (29)$$

This relationship differs from the Navier–Coulomb failure criterion only by the fact that the cohesive strength is replaced by $2T$ and μ_s replaces μ_i.

Since closure of the cracks will not take place in a tensile stress field, the shape of the envelope in tension will be determined by eqn. (25). Consequently, the complete envelopes will be of the form represented in Fig. 16.

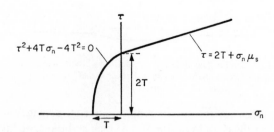

FIG. 16. Relationship between shear and normal stresses at failure showing the curved relationship predicted by the Griffith theory in the tensile stress quadrant and the linear relationship predicted by the modified Griffith theory in the compressive quadrant.

The simple Griffith theory results in a relationship between principal stresses at failure where the constant B in eqn. (17) is approximately equal to 2·0. While if the theory is modified and the cracks and flaws are assumed to close under very low compressive stresses $\sigma_c = 0$, the resulting relationship between principal stresses is linear and $B = 1·0$. Clearly, if it is assumed that cracks close at various levels of compressive stress, it is possible to derive relationships between principal stresses at failure where B may have any value between 1·0 and 2·0.

Moreover, it has been found experimentally that the value of the intermediate principal stress σ_2 does have a small influence on the conditions of failure. The simple Griffith theory has been extended by Murrell (1962), and in triaxial compression it predicts that the intermediate principal stress will affect the conditions of failure, but that the influence is of the second order.

Thus, the modified Griffith theory satisfies the experimental data cited earlier. Also, when the relationship between principal stresses is linear, it puts the Navier–Coulomb criterion on a sound theoretical basis.

CRACK PROPAGATION

It is emphasized that the Griffith theory deals only with the conditions which are necessary to initiate crack propagation, and not with the subsequent development of the propagating crack.

The development can be expected to be straightforward, in tension, for a single crack may grow by lengthening in a plane at right angles to the tensile stress until complete failure occurs.

In compression, the exact mode of development cannot be readily anticipated by theoretical considerations. To overcome this difficulty Brace and Bombolakis resorted to a simple, but ingenious, photoelastic technique. In their experiments, a Griffith crack is simulated by an elliptical hole cut in a sheet of photoelastic material. When loaded, the stress distribution around the flaw, and in particular the points of maximum tensile stress, can be ascertained. If the flaw propagates, it will do so at the points of maximum tensile stress. The propagation of the flaw can be simulated by cutting away the points of maximum tension in the photoelastic model. By re-applying the load, further zones of tension in the model can be established which, in turn, are cut away so that the crack is made to "grow".

Brace and Bombolakis found that "cracks" tended to propagate along a curved path which gradually became parallel with the direction of principal compression. This conclusion was borne out by experiments on glass. Further, it was found that when the crack became parallel with the axis of compressive stress, crack growth ceased. Hence, it follows that a single flaw will not develop into a macroscopic shear failure.

They also considered the interaction of flaws and, in preliminary experiments, found that systems of *en echelon* cracks propagate at a fraction of the applied stress necessary to cause growth of an isolated crack of the same orientation and size.

These experiments and conclusions apply to homogeneous and isotropic material. It may be anticipated that the pattern of behaviour of rock will be more complicated.

Most rocks are not completely isotropic, nor can the cracks be expected to be either elliptical or of equal size. Consequently, because of the diversity in the shape and size of natural flaws, cracks and pore spaces in rock, it can reasonably be assumed that failure of a rock mass will not occur at a single critical pressure. Instead, one may envisage that local microscopic failure will be initiated at a few suitably shaped and orientated flaws. These flaws will propagate for a short distance, probably until they encounter another flaw of different shape, size and orientation, whereupon the local stress concentration will be radically altered and the flaw may then cease to propagate. At higher levels of applied stress, other flaws with a suitable shape and orientation will propagate, so that eventually a network of flaws will develop. Finally, these enlarged flaws will run together and macroscopic shear failure will result.

In an oral communication to the author, Amery stated that the development of these flaws (which he called protoshears) has been observed during a compression test of a specimen which was coated with a photoelastic material.

MICROSEISMS

It has been noted that when failure of a brittle material takes place, a portion of the energy released occurs as vibrations which are frequently audible. For example, the failure of a competent rock specimen at the end of a uniaxial compressive strength test frequently occurs with explosive violence and is accompanied with a noise comparable with that of a gun-shot. However, noise at a much lower level of intensity is also generated during the propagation of micro-flaws, long before macroscopic failure occurs.

The cracking noises (Obert has termed them microseisms) generated by specimens during a compression test can be detected by geophones attached to the specimen and the signals they pick up are suitably amplified. The occurrence of microseisms and the associated formation of the flaws and cracks have been observed in concrete, rock and ice.

TABLE 4

Onset of Microseismic Activity

Rock type	Load (expressed as percentage of failure load) at which micro-seisms first occur (%)
Amygdaloidal Basalt	75
Medium-grained Limestone	27
Epidosite	48
Granite (A)	26
Granite (B)	38

For example, von Rusch recorded the internal cracking which developed in concrete specimens which were subjected to compressive loads. He observed a high rate of microseismic activity for stress levels equivalent to 85 per cent of the short-term failure load. However, the generation of cracks in concrete probably begins at a much lower level of stress, for Jones (1952) observed that the velocity of an ultrasonic pulse transmitted through a concrete specimen began to decrease when the load was approximately 25–30 per cent of the failure load. This decrease can almost certainly be attributed to the formation of internal cracks.

Cracking activity of rock *in situ* during mining operations has been observed by Obert and Duval. They also studied the microseismic activity of a number of rock types subjected to compression in the laboratory. They found that the onset of cracking varied somewhat (see Table 4) but, in general, it began at between 25 and 50 per cent of the failure load obtained in short term tests (i.e. tests with a duration of only a few minutes).

The data regarding the microseismic activity in rock in short duration compression tests is sparse; nevertheless, it is extremely likely that most, if not all, competent rocks begin "cracking up" at a level of stress far below that at which they finally fail.

Further aspects of the experiments on the microseismic activity of rock and of ice will be dealt with in a later section.

Stress–strain experiments (Price, 1958; Hobbs, 1964a) show that during compression tests of short durations, competent and moderately competent rocks, even when loaded to stress levels close to their failure load, usually behave as elastic material. Hence, it may be inferred that the formation and propagation of flaws during such short term compression is essentially a form of micro-brittle failure. Plastic deformation, if it occurs, is of negligible proportions.

It is emphasized that the experimental data cited in previous paragraphs were derived from short term experiments. However, Serdengecti and Boozer have demonstrated that a slow strain rate facilitates ductile deformation, while Griggs (1939) and others have shown that rocks will "creep" and show permanent deformation under stress much lower than the instantaneous failure load, providing the stress is applied for long periods.

Consequently, it is necessary at this point to consider the influence of "time" in the process of rock deformation.

TIME–STRAIN

A study of time–strain, or creep, in a stressed material forms an important part of any rheological investigation. Consequently, there is a considerable amount of data relating to substances ranging from paint and toothpaste to metals, concrete and other engineering materials, available in a large number of standard texts (Finnie and Heller; Eirach, etc.). However, data relating to creep in rock are scarce.

The time–strain pattern exhibited by the wide range of materials mentioned above, subjected to a constant uniaxial stress, is surprisingly similar in form and is represented diagrammatically in Fig. 17. The *instantaneous elastic strain*, which takes place when a load is applied, is represented by *OA*. There follows a period of *primary creep AB* in which the rate of deformation decreases with time. Primary creep is sometimes referred to as *delayed elastic deformation* or *elastic flow*, for if, at time T_1 the specimen is unloaded there is first an instantaneous elastic recovery *BC* followed by a *time-elastic* recovery, represented by curve *CD*.

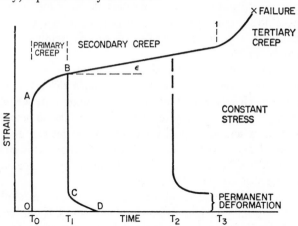

FIG. 17.   Theoretical time–strain curve at constant stress.

However, if the load is not removed at time T_1, the specimen begins to exhibit *secondary creep*, a phase of deformation in which the rate of strain is constant. Because the rate of strain is a constant and because the specimen has undergone a permanent deformation, as would be shown if the specimen were unloaded at time T_2, this deformation is sometimes termed *pseudo-viscous*.

However, if the specimen is not unloaded at time T_2 the rate of deformation increases until the specimen eventually fails. This phase of deformation is known as *tertiary creep*.

The time–strain can be expressed as a function of stress, temperature or other variables, and empirical equations may be fitted to the experimental results. This approach was used by Andrade in his work on tensile creep in metals.

An empirical equation, which has been related to creep of rock in compression, is that presented by Griggs (1939). He suggested that the total strain at any time t up to, but not including, the phase of tertiary creep, can be represented by the empirical equation

$$e_t = A + B \log t + C \cdot t \tag{30}$$

where A, B and C are constants of the material. According to Griggs, A represents the instantaneous elastic component of the strain, $B \log t$ represents the time-elastic, or primary, creep component, and $C \cdot t$ represents the phase of pseudo-viscous, or secondary, creep.

This empirical relationship appears to represent with reasonable accuracy the overall time–strain data obtained from a number of rock types. However, as Griggs himself points out, the $B \log t$ component cannot possibly represent the time-elastic deformation for very large and very small values of t, since this expression will give rise to a positive and negative infinitely large elastic strain for $t = \infty$ and $t = 0$ respectively. The latter difficulty can be obviated if the $\log t$ term is replaced by $\log (t + 1)$.

A theoretical approach has been made in the field of metal physics. By applying the principles of statistical mechanics, it can be shown (Freudenthal) that the viscosity η_a of a homogeneous crystalline material can be expressed as

$$\eta_a = \text{const. exp.} \, (Q/RK^0{}_A) \tag{31}$$

where Q is the heat of activation, R is Boltzmann's constant and $K^0{}_A$ is the absolute temperature. This approach has been used by Serdengecti and Boozer in a study of the strain rate on the behaviour of Solenhofen Limestone subjected to triaxial compression. However, the application of these concepts to time-dependent strain in the majority of rock types, which are heterogeneous and complex in composition and structure, cannot at present be envisaged.

It is often convenient to express the observed behaviour of the material in terms of a rheological "model" which is built up of a combination of three simple isotropic units. These three units can be represented graphically as a spring, a dashpot and a sliding weight (see Fig. 18). The spring on its own represents a simple "Hookean" elastic body; the dashpot on its own represents the viscous behaviour of a "Newtonian" liquid, while the weight represents a "St. Venant" body which simulates the yield strength of a simple plastic solid. These basic units can be combined in series (|) or in parallel (—). Thus, a spring in series with a dashpot (known as a Maxwell model) represents the simplest form of elasticoviscous behaviour, while a spring in parallel with a dashpot (known in combination as a Voigt unit) represents elastic flow behaviour.

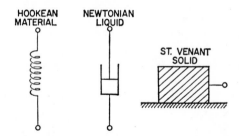

Fig. 18. A spring, dashpot and weight representing, diagrammatically, basic rheological units with specific behaviour patterns.

More complex and sophisticated combinations of units can be postulated to represent the observed strain behaviour of actual materials.

One such model which can be used to interpret time–strain curves, of the type shown in Fig. 17, is made up of a Maxwell unit M coupled in series with a Voigt unit V (see Fig. 19a). Using rheological nomenclature (see Reiner) this model may be termed an M–V body. It is also referred to as a Burger's body.

In this model, the spring E_i represents the component of the body which gives rise to the instantaneous elastic strain. The spring and dashpot coupled in parallel $(\overline{E_t\ \eta_k})$ represents the component responsible for primary, or time-elastic, creep. The component of secondary creep, or pseudo-viscous flow, is contributed by the dashpot η_m.

It can be inferred that this model will eventually exhibit permanent deformation even when the deforming stress becomes exceedingly small, and therefore represents a complex liquid.

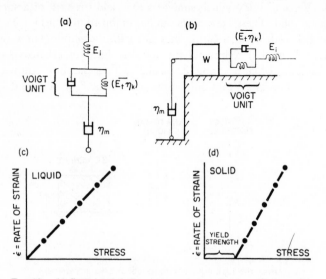

FIG. 19. (a) Spring and dashpot representation of a visco-elastic model. (b) Spring and dashpot representation of B–V model. (c) Relationship between rate of secondary creep and stress for an elastico-viscous body. (d) Relationship between rate of secondary creep and stress for a B–V model.

Providing the stress exceeds the yield strength of the material, the time–strain behaviour indicated in Fig. 17 can be represented by the model shown in Fig. 19b. The dashpot η_m, weight W and spring E_i together form a Bingham body B. Hence, the model is made up of a Bingham body in series with a Voigt unit. This composite model, which has no recognized name, may be referred to as a B–V model.

As in the Burger model, the spring E_i accounts for the instantaneous elastic deformation, and the coupled spring and dashpot $\overline{(E_t\eta_k)}$ accounts for the time-elastic flow. However, before pseudo-viscous, or plastico-viscous, deformation by the dashpot η_m can take place, the frictional resistance of the weight W must be overcome. This frictional resistance, in the model, simulates the long-term strength of a solid.

In order to decide which, if either, of these models represents the time–strain behaviour of a rock it is necessary to obtain creep data for a given rock type at a number of stress levels. The relationship between the rate of secondary creep and stress, which these two model bodies will possess, is indicated in Fig. 19c and d.

Price (1964) conducted a series of time-deflection experiments on beams of Pennant and Wolstanton Sandstones in a simple bending beam apparatus. The results obtained are shown in Fig. 20a and b.

It will be seen that the time-deflection curve for beam I of Pennant Sandstone exhibited primary, secondary and tertiary creep and finally failed. The remaining five specimens showed only primary and secondary creep. The instantaneous elastic deflection is not shown in these diagrams.

The time-deflection curves for Wolstanton Sandstone are shown in Fig. 20c. Specimens submitted to loads ranging from 65 to 85 per cent of the instantaneous failure load show both primary and secondary creep. However, the specimen submitted to 50 per cent of the instantaneous failure load exhibits no evidence of secondary creep.

These time-deflection data, which are almost completely the result of creep in extension, permit the relationship between load and the rate of secondary creep to be established. It will be seen from the inset diagrams that there is, in fact, a linear relationship between these two parameters and that the line gives abscissae on the load axis. This is in good agreement with the rheological principles noted earlier and indicates that these rock types have, in extension, the behaviour pattern of a B–V model and not that of a visco-elastic model.

It will be seen that the long-term strengths of these rocks in extension are approximately 20 and 60 per cent of the instantaneous strength for Pennant and Wolstanton Sandstones respectively.

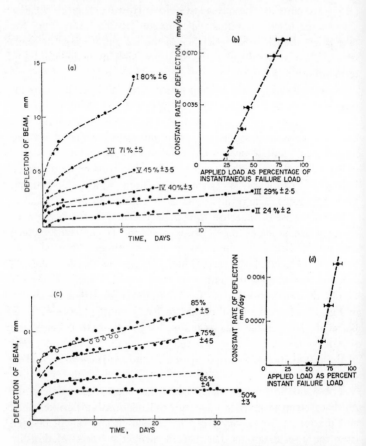

FIG. 20. (a) Time–deflection data for beams of Pennant Sandstone. (b) Relationship between load and corresponding constant rate of deflection. (c) Time–deflection data for beams of Wolstanton Sandstone. (d) Relationship between load and corresponding rate of deflection.

Misra has conducted creep experiments in compression on a number of igneous and sedimentary rocks. These experiments were of relatively short duration, for they do not exceed 14 days. However, they were conducted at temperatures ranging from 20 to 475°C.

The data obtained were used by Misra to obtain values for the energy of activation of the various materials [i.e. he determined the value of Q in eqn. (31)]. The time–strain curves he obtained may also be treated in the manner indicated in Fig. 20 and in the previous paragraphs. Details of the long-term strength, as percentages of the instantaneous strength, for the various rock types he tested are listed in Table 5.

TABLE 5

Rock type	Long-term strength as percentage of instantaneous strength	
	%	
Dolomite	50	at 20°C
Limestone	35	at 20°C
Darley Dale Sandstone	50	at 20°C
Granodiorite	27	at 20°C
Darley Dale Sandstone	50	at 400°C
Granodiorite	22	at 475°C

The rate of creep of the Darley Dale Sandstone and the Granodiorite are increased at the higher temperatures, but it can be seen that the long-term strength of the Sandstone is unaffected by temperature, while the strength of the Granodiorite is reduced by only 5 per cent.

Griggs (1940) conducted a similar series of experiments on "wet" alabaster, for periods up to 100 days, and obtained primary, secondary and tertiary creep curves at stresses ranging from 1500 to 4250 lb/in². An analysis of his results indicates that even in conditions conducive to recrystallization the long-term strength of alabaster is approximately 30 per cent of its instantaneous strength.

From these various data it would appear that the long-term strength of rock in uniaxial compression and tension are from 20 to 60 per cent of their instantaneous strength.

However, a criticism which may be levelled against these, and indeed against most creep experiments when applying the conclusions to geological problems, is that the duration of the test is so short in comparison with the length of time involved in most geological processes. Thus, whereas the B–V model may represent the behaviour of rocks for engineering problems, where the duration will be measured in tens of years, there exists considerable doubt whether this model will suffice for geological problems where durations may be measured in terms of millions of years. Indeed, it has been noted that there exists an "atomic viscosity" η_a which may be interpolated in the B–V model resulting in the hybrid model represented in Fig. 21. If it is assumed that η_a is very much larger than η_m, this model will behave as a simple B–V model for periods of a few tens of years, but may behave as an elastico-viscous model when the stresses are applied for thousands or millions, of years. Hence, the question whether rocks in the upper layers of the crust are to be considered as liquids or solids, is still unanswered.

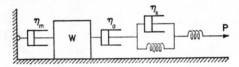

Fig. 21. Diagrammatic representation of B–V model including an "atomic" dashpot η_a.

Diagnostic laboratory evidence presented itself, quite fortuitously, during the investigation of creep in compression of a nodular, muddy limestone (Price, 1964). The samples from which the test specimens were prepared were obtained from a mine and were completely unweathered. Hence, to ensure that the specimens were not undergoing strain as a result of the extraction from the mine, they were placed in position on the loading rig, with strain measuring device attached, in an unloaded state for a period of 45 days. The behaviour of only one of the specimens need be considered here. After the initial 45 days, it was submitted to a compressive load of 5250 lb/in², which was maintained constant for 70 days. At the instant the load was applied, the specimen exhibited instantaneous elastic shortening (see Fig. 22). But almost immediately after the load was applied, the specimen began to expand. The rate of expansion was relatively rapid at first, but after the first 7 days it settled down to approximately 0·7 micro-strains (μs) per day. When, after a period of 70 days, the specimen was unloaded, it expanded, showing instantaneous and time-elastic recovery, so that after a total period of 140 days it was approximately 350 μs longer than it had been originally. At this time the specimen was again subjected to a compressive load of 5250 lb/in². It again showed instantaneous elastic shortening, but instead of expanding as it had in the previous cycle of the test, the specimen showed "normal" primary and secondary creep. When the specimen was subsequently off-loaded and elastic recovery was complete, it was some 80 μs shorter than it had been immediately prior to the second application of load.

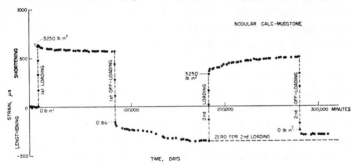

Fig. 22. Time–strain data obtained from specimen of nodular muddy limestone showing expansion during and after the first loading and normal time–strain behaviour during the second loading cycle.

It is emphasized that the experimental procedure in both loading cycles was identical. Consequently, the difference in the time–strain data must be attributed to some fundamental change in the test specimen.

The writer has suggested that there was a release of strain energy stored in the rock specimen during and after the first period of loading, and it was this release of energy which brought about the increase in dimension of the specimen.

The mechanism envisaged is as follows. It may be inferred from the spring and dashpot model, represented in Fig. 19a, that if a Burger body is maintained in a state of constant strain, the elastic strain energy needed to deform the model is initially stored in the spring, but over a period, the energy is dissipated by bringing about viscous flow in the dashpot η_m. The proportion of energy dissipated in this way is determined by the length of time during which the constant strain conditions are maintained relative to the relaxation time of the model. The relaxation time t_R, which is the time required for the stored stresses associated with the strain energy to reduce to $1/e$ of their initial value, is given by

$$t_R = \eta_m/G \qquad (32)$$

where η_m is the viscosity in poises and G is the modulus of rigidity in dynes per square centimetre. For this particular rock type, assuming it to behave as an elastico-viscous model, the relaxation time would be approximately one year.

If a B–V model is maintained in a condition of constant strain, only a portion of the strain energy is dissipated by internal viscous flow. For, once the stress falls to a level equal to the yield strength of the model, all internal viscous flow ceases, i.e. the spring is unable to move the weight W in Fig. 19b and an elastic stress equal to the yield strength is held in the model for an indefinite period, as long, in fact, as the constant strain conditions are maintained.

If rocks behaved in a manner comparable with this simple model, all strain energy would be dissipated from a rock sample within a short time of its extraction from the solid. However, it will be recalled that during the test on the nodular, muddy limestone, care was taken to ensure that the specimen was not expanding prior to being placed on load. Hence, the subsequent expansion of the specimen still poses a problem.

It is suggested that this difficulty can be resolved, for whereas the B–V model is assumed to represent a homogeneous and isotropic material, the rock type under discussion was neither homogeneous nor isotropic. Consequently, if the rheological model is brought into closer agreement with the observed composition and structure of rock by assuming that each individual component particle has the properties of a simple B–V model, then the rock as a whole, because of its cohesion, will behave in a manner compatible with the observed test data. Thus, postulating such a complex model, in view of its heterogeneity, the distribution of stresses in the rock on the microscopic scale can never be uniform. Consequently, when a rock sample is extracted from the solid only a portion of the strain energy is dissipated by expansion. However, a sedimentary rock, in its geological history, has been compacted, has had its pore spaces filled and its grain boundaries fused or cemented, so that the problem is not one of simple elasticity. Hence, even after this expansion, it is possible that some grains, or groups of grains, may retain a certain amount of residual strain energy. If the cohesion between the strained components is broken down, a further change in dimension of the specimen will result from the dissipation of residual strain energy. It is suggested that such a breakdown is brought about when the specimen is loaded during a creep test.

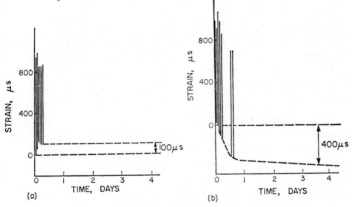

Fig. 23. (a) "Pseudo-plastic" shortening after "pulse loading" of a specimen of nodular, muddy limestone. (b) Expansion of rock specimen after pulse loading.

To check the validity of this suggestion, experiments were conducted in which specimens from Warsop Colliery were pulse loaded and the subsequent time–strain of the unloaded specimen was observed. A single pulse was of less than 5 minutes duration, and was the time required to build up the load to 5250 lb/in² and reduce it again to zero. In this short period, the strain due to creep was negligibly small so that the subsequent time–strain can, with little doubt, be related to the residual strain-energy of the specimen. The data obtained from two such experiments is shown in Fig. 23.

It is reasonable to assume that the residual strain-energy in the specimen, which shortened in a pseudo-plastic manner, is predominantly tensile, while for the specimen that expanded, the increase in length can be attributed to the release of compressive strain-energy. It is suggested that, although the breakdown of internal strained boundaries was instantaneous because the energy thus released has to work out of the specimen by straining the surrounding material, the recovery will not, in general, be instantaneous but will occur as a time–strain effect as shown in Fig. 23.

It may be inferred from the data in Figs. 22 and 23 that, if the elastic stress–strain relationship of this rock type is reasonably linear, the average levels of residual stress (associated with strain energy) which gave rise to an expansion of 350 μs and 400 μs were approximately 3000 lb/in² and 2000 lb/in² respectively. However, the mechanism proposed to explain the retention of these residual stresses requires that they be localized in pockets and not distributed evenly throughout the specimen. Hence, the level of stress in these pockets is probably several times greater than the estimated average value of 2–3000 lb/in². It is tentatively suggested that a value of not less than 10,000 lb/in² for these pockets of residual stress would be a reasonable estimate.

The rock sample from which these specimens were prepared was obtained from mine-workings at a depth of 1600 ft. Thus, the estimated values for the residual stresses are greater than the values one would expect to be generated at this depth as a result of the gravitational loading. Consequently, it may be concluded that the pockets of strain energy with their associated stresses are geological in origin and are, therefore, of considerable antiquity.

It will be recalled that a visco-elastic material with the same physical constants as the nodular limestone would have a relaxation time of only a year. Therefore, in the light of the conclusion of the previous paragraph, it is quite clear that this rock type is not an elastico-viscous material.

However, these experimental results do not completely obviate the possibility that, if such rocks are subjected to low differential loads for an infinite time, they may yet behave as a liquid of high viscosity (see Fig. 21). Since these experiments demonstrate that the specimen has retained residual stresses which were of great age, it follows that the relaxation time of the body as a whole must be many tens or even hundreds of millions of years. Consequently, η_a must have a viscosity in excess of 10^{26} poises. Such an atomic viscosity would have little practical significance even in geological events of long duration.

As with many problems relating to rock mechanics, the experimental data are scarce. Nevertheless, the data dealt with are representative of a large number of competent rocks. Consequently, it is reasonable to assume that other rocks of a similar type with similar geological histories will also behave as B–V models and, moreover, that a high percentage of these will have a long-term uniaxial strength of one quarter to one half of their instantaneous strength.

There are two points which need emphasis.

Firstly, the total amount of time–strain exhibited by specimens in uniaxial compression and bending tests is very small, i.e. less than 500 μs, or 0·05 per cent. This amount of time–strain is vastly different from the 1–25 per cent frequently recorded in creep experiments in metals at elevated temperatures and pressures, or the 14 per cent strains obtained in specimens of Solenhofen Limestone subjected to a confining pressure of 10,000 atmospheres and a differential pressure of almost 100,000 lb/in^2 (Griggs, 1936). The mechanism involved in the deformation is also likely to be different.

Secondly, the stress level at which failure takes place depends upon the length of time for which the stress is applied. Thus, it will be noted that specimen I in Fig. 20a was subjected to a load of 80 per cent of the instantaneous failure load, and was able to sustain this load for almost a week before it eventually failed. The total time-deflection of this beam was approximately equal to the instantaneous elastic deflection the beam would have experienced had the applied load been increased from 80 to 100 per cent.

The influence of time, or in other words, the rate of deformation, is not considered by the static criteria of brittle failure of Navier–Coulomb and Griffith. A theory which does take the rate of deformation into account is that due to Reiner and Weissenberg (see Reiner). They consider that failure of material takes place when the *strain-work* has reached a certain limit; where strain-work is defined as the conserved part of all the work done by the applied stress (a portion of which will be dissipated in heat, vibrations, etc.).

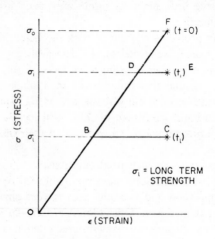

Fig. 24.   Hypothetical stress–strain relationship to failure based on Reiner–Weissenberg concept of failure.

It may tentatively be inferred from this concept, the rheological model represented in Fig. 21, and the time-deflection data for beam I in Fig. 20a, that the stress-strain relationship prior to failure depends upon the intensity of the applied stress and the duration of its application in the manner indicated in Fig. 24. If the stress σ_0 is the instantaneous failure load, the line OF represents the instantaneous stress–strain curve leading to failure at point F when the load is

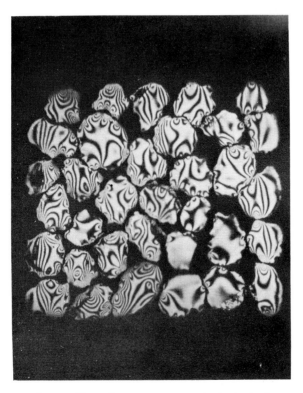

Plate 1. Photo elastic stress concentration in simulated
granular material submitted to uniaxial compression.

Plate 2. Joint surfaces in Solenhofen Limestone showing "augen", or rib, structures and also faint plumose markings. (Photo reproduced by courtesy of I. Gramberg *et al*)

applied "instantaneously" (i.e. $t = 0$). If a stress σ_l equal to the long-term uniaxial strength of the specimen is applied, the elastic strain will be given by OB on ϵ axis. If this stress is maintained for a very long, but not infinite, time, the time–strain is represented by BC with failure finally occurring at point C. For a stress σ_i, intermediate between σ_0 and σ_l, the instantaneous strain is given by OD and the time–strain by DE, with failure taking place at point E at some time t_c where $t_l > t_i > t_0$. Thus, to produce failure, providing the applied stress is greater than σ_l, one may "feed into" the specimen a lot of energy for a short period, or a little energy for a long period.

It is pointed out that the limit of σ_l cannot be rigorously defined. At infinite time σ_l, theoretically, must equal zero. But this is of little practical importance, for the experimental data presented earlier indicate that σ_l is a significant proportion of the short-term strength for periods measurable in millions of years.

TIME–STRAIN MECHANISM IN SEMI-BRITTLE MATERIAL

Creep in a crystalline solid can result from one, or a combination, of a number of mechanisms which include deformation by differential slip along crystal glide planes, by twinning, plastic or viscous deformation of the crystal lattice, or by the propagation of flaws, etc., and development of non-elastic differential movement along such flaws, cracks and dislocations.

Deformation by crystal gliding and twinning is important in metals and in the carbonate minerals, especially at elevated temperatures and pressures. These particular mechanisms appear to be of negligible importance in the deformation of quartz at low temperatures and confining pressures. It may be inferred from the lack of evidence of an increase in the amount of strain shadows in quartz grains in creep specimens following a test, that general deformation of the crystal lattice is also of negligible importance at these low levels of temperature and pressure. By elimination, the remaining mechanism is that of crack and flaw propagation accompanied by a small amount of non-recoverable differential movement.

In addition, there is positive evidence which indicates that micro-cracking and flaw propagation occurs during creep. For example, the release of pockets of residual strain-energy in a specimen, described earlier, has been attributed to the formation of micro-cracks. It will be recalled that the microseismic activity recorded by Obert and Duval began when the applied stress was from one quarter to one half of the instantaneous failure strength. It has been noted that the long-term strength of many rocks is also likely to be between one quarter and one half of the instantaneous failure strength. The author feels that these relationships are highly significant and suggests that the onset of microseismic activity probably occurs when the stress applied to the specimen exceeds the long-term strength of the test specimen.

In addition, Obert and Duval found that if the applied load was maintained constant for a period, at various stress levels up to 90 per cent of the instantaneous failure load, the initial high rate of microseismic activity decreased with time, eventually reaching a constant level of activity which was related to the magnitude of the applied load. It is tempting to correlate these variations in the micro-seismic activities with primary and secondary creep in the specimen. Gold has carried out similar tests with ice, and, because the trans-parency of ice is marred by the formation of flaws, he was able to demonstrate conclusively the importance of crack formation and propagation in the creep process.

As we have seen, the Griffith concept of flaw propagation in a brittle material depends on the development of extremely high local stresses around a micro-flaw which, if failure occurs, causes atomic separation and propagation of the flaw. If the applied stresses are not sufficiently high, brittle failure will not take place in the ideal homogeneous and isotropic material which was postulated by Griffith. Nevertheless, there will exist around the flaws certain areas, in which the tensile stresses are exceedingly high (possibly as great as $10^6 \mathrm{lb/in^2}$).

Now the heat of activation of the atoms Q is influenced by such a local stress field σ and is given by

$$Q_\sigma = (Q - b \cdot \sigma) \tag{33}$$

where b is a constant. Hence, the atomic viscosity η_a is locally reduced [see eqn. (31)]. Moreover, the stress will, in general, tend to distort the space lattice, strain will be inhomogeneous, so that plastic or slow viscous dislocation and differential movement will be promoted.

From the experimental time–strain data and the concept of the previous paragraph it may be inferred that, provided the applied stress is greater than some critical value which is equal to the long-term strength of the material, flaws will propagate. Some suitably orientated and shaped flaws will propagate relatively soon after the load is applied and will be accompanied by a negligible amount of plastic deformation. It is suggested that the propagation of such flaws is responsible for the essentially elastic phase of primary creep. Less favourably orientated and shaped flaws are associated with less intense micro-stress fields. Such flaws will propagate less readily and will be accompanied by a time-delay and a proportionally greater amount of plastico-viscous deformation. The propagation of these flaws could account for the phase of secondary creep. As the network of flaws builds up and individual flaws increase in size and inter-connect, deformation becomes progressively more rapid (tertiary creep), until a network of flaws, comparable with that which causes instantaneous brittle failure, is established and long-term semi-brittle shear failure occurs.

Rocks are not completely homogeneous and isotropic, so that any one specimen is unlikely to possess a completely uniform strength throughout the whole specimen. In certain small portions, the strength may be less than the long-term strength of the specimen as a whole. It is suggested that a limited amount of crack development will take place in these weaker portions, and that this localized development may account for the occurrence of primary creep even at stress levels below the long-term strength of the specimen as a whole.

The condition for instability of a fatigue crack, which spreads as a result of plastic deformation, has been established by McClintock and he has shown that the elastic-plastic analysis approaches that of the Griffith brittle failure criterion.

LONG-TERM STRENGTH IN TRIAXIAL COMPRESSION

From the experimental evidence and the intuitive description of the possible time–strain mechanism given above, it seems likely that a modified form of the Griffith criterion will, as a close approximation, also serve for a long-term failure criterion.

It will be recalled that the simple Griffith criterion gives rise to a relationship between principal stresses at failure which is expressed in terms of stress parameters and a single physical constant, the instantaneous tensile strength T. It is tentatively suggested that if the long-term tensile strength T_l is substituted for T in eqn. (25) the long-term strength of rocks with markedly non-linear relationships between principal stresses at failure will approximate to

$$(\sigma_1 - \sigma_3)^2 + 8T_l(\sigma_1 + \sigma_3) = 0. \tag{34}$$

Further, if it is assumed that the coefficient of sliding friction is unaffected by time, the long-term strength in triaxial compression for those rocks with a linear relationship between principal stresses at failure will approximate to

$$\sigma_1 = \sigma_l + K\sigma_3 \tag{35}$$

where σ_l is the long-term uniaxial compressive strength and has a probable value of one quarter to one half of the instantaneous strength.

It is readily admitted that the above suggestions regarding the modification of the equations based upon the Griffith criterion is contentious and open to debate. Clearly much practical and theoretical work will be necessary before such a hypothesis can be established or, alternatively, shown to be unsound. The relatively rapid increase in strength with increase in confining pressure indicated by eqns. (34) and (35) will probably result in an over estimate of the strength. Nevertheless, in the opinion of the author, it provides the most reasonable means at present available of estimating the long-term strength of competent rock in a stress and temperature environment which is conducive to an essentially brittle, albeit delayed, form of fracture.

CHAPTER 2

FAULTS

INTRODUCTION

A fault is a plane of fracture which exhibits obvious signs of differential movement of the rock mass on either side of the plane. Faults are, therefore, planes of shear failure. Such shear fractures in competent rock have frequently been interpreted as manifestations of brittle failure. As we have seen in Chapter 1, it may be taken that competent rocks in the upper zone of the crust, where temperatures and pressures are relatively low, behave as brittle or semi-brittle solids. Hence, for these rock types one may assume that the stresses prior to failure may be ascertained with reasonable accuracy by elastic theory. The conditions leading to the initiation of faults can be determined by applying one or other of the criteria of brittle failure; while movement subsequent to failure can be interpreted in terms of frictional sliding of block upon block.

In this chapter, for the sake of simplicity, the development of faults in homogeneous, isotropic rock is dealt with using the linear Navier–Coulomb criterion of failure. This analysis leads to the dynamic interpretation of faults proposed by Anderson. The influence of anisotropy on shear failure is considered briefly and this is followed by a discussion of the formation of certain types of faults and ancillary structures. Methods of stress analysis in elastic systems prior to failure are also dealt with. First, however, it is apposite to deal briefly with the nomenclature used to describe faults.

FAULT NOMENCLATURE

The direction of differential movement along a fault plane can be referred to the orientation of the fault plane itself, so that the movement can be classified as *dip-slip*, *strike-slip* or *oblique-slip*, according to whether the direction of movement is parallel to the dip or strike of the fault plane, or is oblique to these directions. If a fault is inclined to the vertical, the *hanging-wall* is that face of rock which lies above the fault plane while the *foot-wall* is that which lies below. If the hanging-wall moves downward in a dip-slip direction with respect to the foot-wall, the fault is classified as a *normal fault*. If the hanging-wall moves upward with respect to the foot-wall (again in a dip-slip direction) the structure is known as a *thrust* when the dip of the plane is less than 45°, and a *reverse fault* when the dip is greater than 45°. There are, in addition, certain major thrust-like structures where the fault plane dips at less than 10°. Such structures are termed *overthrusts* and, as will be shown later, are best put in a separate, special category of faults.

Fault planes which exhibit strike-slip are commonly vertical, or approximately vertical, and are classified as *wrench* or *transcurrent faults*. If an observer stands facing a wrench fault and the rock mass on the opposite side from the observer has been displaced to the right relative to the observer, the differential movement along the fault plane is variously described as *right-handed*, *dextral*, or *clockwise movement*. Similarly, the opposite sense of movement has been termed *left-handed*, *sinstral* or *anticlockwise movement*.

For oblique-slip faults it is found necessary to combine the above nomenclature and, for example, refer to a left-handed, oblique-slip normal fault.†

THE NAVIER–COULOMB CRITERION AND THE ANGLE OF FRACTURE

In many areas, particularly in those which have not undergone intense tectonic deformation, it is frequently found that normal faults are inclined at angles greater than 45°, while thrust planes consistently dip at angles considerably less than 45°. For example, Sax, who studied the orientation of over 2000 faults in the coal measure rocks of the Netherlands, of which approximately 80 per

---

† For a more detailed classification of faults the reader is referred to the Report of the Committee on the nomenclature of faults, *Bull. Geol. Soc. Am.* **24**, 1913.

cent were normal faults, demonstrated that the most usual inclination of these planes was between 60° and 65° for the normal faults and between 20° and 25° for thrusts. Thus, the acute angle separating complementary sets of such fractures would most frequently be from 50° to 60° for normal faults and 40° to 50° for thrusts. Similar values were obtained by Price (1962) for the inclination and acute angle between sets of normal faults developed in the Aberystwyth Grits. Somewhat higher values are quoted by Lieth.

The acute angle which exists between complementary fault systems is, in fact, predicted by the Navier–Coulomb criterion of brittle failure which, it will be recalled, is given by

$$\tau = S + \mu_i \cdot \sigma_n.$$

It can be shown that this criterion of failure leads to the development of shear planes which intersect to form an acute angle in the following manner.

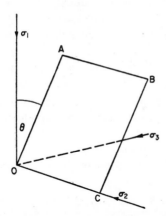

FIG. 25. Inclined plane in a triaxial stress field with dip direction of the plane making an angle of θ with the vertical axis of greatest principal stress σ_1.

Consider a plane $OABC$ (see Fig. 25) in a triaxial stress field which is orientated so that the plane is parallel to the axis of intermediate principal stress and makes an angle θ with the axis of greatest principal stress in the σ_1–σ_3 plane. The shear stresses on the plane are due only to the greatest and least principal stresses and are given by

$$\tau = \frac{\sigma_1 - \sigma_3}{2} \cdot \sin 2\theta. \qquad (7)$$

Clearly, the maximum shear stress occurs when $\sin 2\theta = 1$. This occurs when $2\theta = 90°$. Hence, the maximum shearing stress occurs when the plane $OABC$ makes an angle of $45°$ with the axis of the maximum principal stress.

The normal stress acting on the plane is given by

$$\sigma_n = \frac{\sigma_1 + \sigma_3}{2} - \frac{\sigma_1 - \sigma_3}{2} \cdot \cos 2\theta. \qquad (6)$$

Thus, σ_n is also a maximum when $\theta = 45°$.

At values of $\theta < 45°$ the values of τ and σ_n decrease, but the rate of decrease of σ_n, as θ becomes smaller, is greater than the rate of decrease of τ. The optimum conditions of shear according to the Navier–Coulomb criterion can be ascertained by substituting eqns. (6) and (7) in eqn. (18), and determining the position for which

$$\frac{\sigma_1 - \sigma_3}{2} \cdot \sin 2\theta - \mu_i \left(\frac{\sigma_1 + \sigma_3}{2} - \frac{\sigma_1 - \sigma_3}{2} \cdot \cos 2\theta \right) \qquad (36)$$

is a minimum.

By differentiating this quantity and equating the result to zero one obtains the result that the optimum conditions for shear occur when

$$\tan 2\theta = 1/\mu_i . \qquad (37)$$

Thus, the angle which the plane of fracture will make with the axis of principal stress is given by

$$\theta = 45° - \phi_i/2 \qquad (38)$$

where $\tan \phi_i = \mu_i$.

If conjugate planes develop in isotropic material, the acute angle 2θ will be bisected by the axis of maximum principal stress, and the line of intersection of the shear planes will be parallel to the axis of intermediate principal stress.

The angle which the shear planes will make with the axis of greatest principal stress for a few representative values of μ_i are given on page 61:

$$\theta = \pm 45° \qquad \text{when } \mu_i = 0,$$
$$\theta = \pm 30° \qquad \text{when } \mu_i = 0.57,$$
$$\theta = \pm 22.5° \qquad \text{when } \mu_i = 1.00.$$

The angle of fracture may readily be obtained from experimental data by the construction of the Mohr's envelope, for the slope of the envelope is equal to φ_i and, consequently, the angle included by a normal to the envelope (NN' in Fig. 11) and the σ-axis is $90° - \phi_i$ and equals 2θ. For a linear envelope the angle of fracture is obviously a constant.

This type of construction may also be used to determine the angles of fracture when the envelope is curved. In such instances, 2θ for any particular stress condition is that angle included between the σ-axis and the normal to the tangent of the envelope. This angle is clearly not constant in value for all values of stress, but increases as the confining pressure σ_3 increases.

For homogeneous and isotropic rocks, the angle of fracture, as determined from measurements taken from sheared specimens, is in very good agreement with that which may be predicted from the Mohr construction. For example, the calculated and measured angles of fracture for Pennant Sandstone, a rock type with a linear envelope, are $2\theta = 46°$ and $45°$ respectively; while for Darley Dale Sandstone, a rock with a curved envelope, the measured and predicted angles range from $2\theta = 45°$ to $65°$ and $2\theta = 40°$ to $60°$ respectively (Price, 1958).

DYNAMIC CLASSIFICATION OF FAULTS

Systems of faults and dykes which are formed during one geological epoch commonly have a more or less parallel alignment extending, in some cases, over hundreds of miles. Consequently, it may be inferred that stress systems over large areas of the crust often have a general uniformity in both direction and intensity.

Now it is clear, that no shear stresses can exist at the surface of the earth which act parallel to the surface. Hence it follows, that in the immediate vicinity of the earth's surface one axis of principal stress is normal to the earth's surface, although at the surface itself the intensity of this principal stress is reduced to that of the atmospheric pressure. In view of the large pressures which obtain at depths in the crust, the stress normal to the surface can, therefore, be taken as zero.

In hilly terrain the normals to the surface are not generally vertical, so that there will exist near the surface a zone in which the principal stresses are frequently inclined to the vertical. However, in areas where the topography is other than Alpine in relief, Anderson (1951) in his classic book has suggested that one can assume, with little error, that one axis of principal stress is, in fact, very close to the vertical and, consequently, the other axes of principal stress will approximate closely to the horizontal.

Anderson introduced the concept of the *standard state* in which the horizontal pressure increases in step with the vertical pressure, so that they are everywhere equal. This standard state, which is similar to a hydrostatic pressure, is, as we shall see later, a useful mathematical concept, and may possibly exist in sediments which have consolidated and compacted purely as a result of gravitational loading.

The orientation of the various types of faults may then be explained by assuming that the standard state is modified in one of the following three ways:

1. There is an increase in pressure in all horizontal directions.
2. There is a decrease in pressure in all horizontal directions.
3. There is an increase of pressure in one horizontal direction coupled with a decrease in pressure in the horizontal direction at right angles.

Consider the first of these cases. In general, the stresses are unlikely to be exactly equal, so that one of the horizontal stresses will be the maximum principal stress, while the other, at right angles, will be the intermediate principal stress. If the deviation from the standard state is sufficiently large, the postulated stress system will give rise to the development of thrusts (see Fig. 26a). If the angle of internal friction φ_i is assumed to have the reasonable value of 30°, the faults will dip at 30°. This is somewhat larger than the value quoted by Sax. The possible reason for this is discussed later.

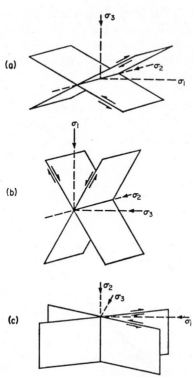

FIG. 26. Relationship of faults to axes of principal stress (after Anderson).
(a) Thrusts—formed when greatest principal stress is horizontal and the
least principal stress is vertical. (b) Normal faults—formed when greatest
principal stress acts vertically and least principal stress is horizontal.
(c) Wrench faults—formed when greatest and least principal stresses act
in the horizontal plane and the intermediate principal stress acts vertically.

In the second case, when there is a relief of pressure in all
horizontal directions, the greatest principal stress acts in the vertical
direction. Again, it is likely that one of the horizontal stresses will be
larger than the other. This system of stresses will give rise to normal
faults (as indicated in Fig. 26b). If, as before, $\varphi_i = 30°$, the faults
will dip at 60°. In this instance, the agreement with the figure given
by Sax and others is quite good.

The third case is that in which there is an increase in pressure in one horizontal direction and a relief in pressure in the horizontal direction at right angles. In this instance, the intermediate principal stress acts in the vertical direction. The wrench faults which can develop as a result of such a stress system are indicated in Fig. 26c.

ANISOTROPY AND SHEAR FAILURE

It has been found in laboratory experiments that when rock specimens possess a well-defined lamination, the angle of fracture, as measured on sheared specimens, is often significantly different from the angle of fracture calculated from the Mohr construction.

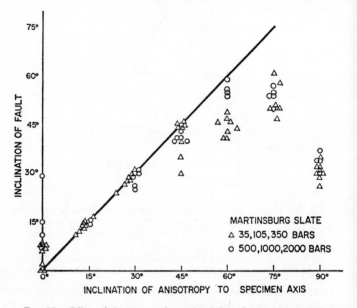

FIG. 27. Effect of cleavage on the angle of shear fracture, Martinsburg Slate (after Donarth).

The influence the orientation of such laminations, or planes of anisotropy, has upon the shear angle has been demonstrated by Donarth. The data he obtained from specimens of Martinsburg Slate are given in Fig. 27. It will be seen that when the cleavage planes are at 90° to the axis of the specimen and, therefore, at right angles to the direction of compression, the angle of shear fracture is 30°. In this instance, the effect of the anisotropy in the specimen is zero. However, when the cleavage is orientated at 30° or less to the maximun principal stress, shear failure takes place along the cleavage planes, and even when the cleavage planes make angles of 60° and 75° to the axis of the specimen, it will be seen that the cleavage profoundly influences the angle of shear.

The simplest analytical form of anisotropy is that of a single plane, or group of parallel planes, of weakness, which cut an otherwise isotropic material. For example, joints, faults or cleavage may represent such weakness planes. The cohesion across a simple fracture may well be zero. However, in many instances the plane of weakness will possess a definite cohesive strength, as, for example, when the plane is lined or filled with quartz, calcite or some cementing material. But, in general, the cohesion across such a weakness plane may be expected to be a fraction of the cohesion of the isotropic material in which it occurs.

There is evidence to show that the coefficient of sliding friction is sometimes a little smaller than the coefficient of internal friction (Jaeger, 1959; Pomeroy and Brown). Assuming that the isotropic rock material and the planes of weakness obey the Navier–Coulomb criterion of failure and that the planes of weakness dip in the plane of the greatest and least principal stresses, then the failure conditions may be represented by the Mohr envelopes shown in Fig. 28. Envelope DE represents the conditions of failure of the isotropic material with a cohesive strength C_1 and angle of internal friction φ_i. Envelope BC represents the conditions of failure along a plane of weakness with a cohesive strength of C_2 (where $C_1 > C_2$) and angle of sliding friction φ_s (where $\varphi_i > \varphi_s$). The conditions of failure along a plane of zero cohesive strength and angle of sliding friction φ_s is represented by envelope OA.

For a given confining pressure σ_3, represented by OF, shearing movement will take place along weakness planes with cohesive strengths zero and C_2, which are orientated at an angle $(45° - \varphi_s/2)$

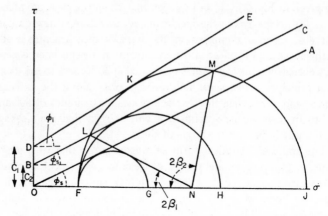

Fig. 28. Diagram showing linear Mohr's envelope for (a) homogeneous isotropic material DE, (b) planar weakness with reduced cohesive strength BC, and planar weakness with zero cohesive strength OA.

to the axis of greatest principal stress, when σ_1 equals OG and OH respectively. If the maximum principal stress attains a value represented by OJ, shear failure may take place either in the isotropic material or along the plane of weakness, depending upon the orientation of the weakness plane with respect to the axis of greatest principal stress.

Consider a plane of weakness with a cohesive strength of C_2. If the angle β which the plane of weakness makes with the axis of greatest principal stress lies in the arc FL or MJ measured clockwise from the point N, the shear stress will be too small to cause movement to take place along the plane of weakness. Instead, a fresh fracture will develop in the isotropic material at an angle of $45° - \varphi_i/2$ to the axis of greatest principal stress. However, if the angle 2β lies in the arc LKM, movement along the plane of weakness is possible and shear failure will not then take place in the isotropic material.

Jaeger (1960) has considered the effect of a more general type of anistropy upon shear failure, by extending the two dimensional theory of the Navier–Coulomb criterion. In the more general case, he assumes that the planar anisotropy of the material can be represented by expressing the cohesive strength as a continuous variable according to the relationship

$$C = a - b \cdot \cos 2(a - \beta) \qquad (39)$$

where a and b are constants and where a and β are assumed to lie between $0°$ and $90°$. Hence, the cohesive strength has a minimum value of $a-b$ when $a = \beta$ and a maximum value of $a + b$ when the plane of anisotropy is rotated through a further $90°$.

Using this expression for the cohesive strength, the conditions of failure become

$$\tau = a - b \cos 2(a - \beta) + \mu \sigma_n. \qquad (40)$$

Now let $(\sigma_1 - \sigma_3)/2 = \tau_m$, and $(\sigma_1 + \sigma_3)/2 = \sigma_m$. Then from eqns. (6) and (7), the stresses τ and σ_n on a plane inclined at an angle a to the axis of greatest principal stress are given by

$$\tau = \tau_m \cdot \sin 2a$$

and

$$\sigma_n = \sigma_m - \tau_m \cdot \cos 2a.$$

Substituting these for τ and σ_n in eqn. (40), the conditions of failure become

$$(\tau_m + b \sin 2\beta) \sin 2a + (\mu\tau_m + b \cdot \cos 2\beta) \cos 2a = a + \mu\sigma_m. \quad (41)$$

For any given value of τ_m the left-hand side of eqn. (41) has its maximum value when

$$\tan 2a = \frac{\tau_m + b \sin 2\beta}{\mu \tau_m + b \cos 2\beta} \qquad (42)$$

The validity of the assumption represented by eqn. (39), for one particular rock type, has been demonstrated by Donarth. From his experimental data obtained using specimens of Martinsburg Slate, it was established that $\mu = 0.5$ and $a = 30°$ and that C varied from approximately 5 to 720 bars (i.e. approximately 70 lb/in² to 10,000 lb/in²) depending upon the orientation of the plane of anisotropy relative to the axis of greatest principal stress. Also, it was established that for a confining pressure of 350 bars (approximately

5000 lb/in^2) the relationship between the greatest principal stress at failure and the angle β is given by

$$\sigma_1 = 226 - 149 \cdot \cos 2 (30° - \beta). \qquad (43)$$

Donarth then expressed eqn. (40) in terms of the principal stresses, and solved for a and b. The values obtained were 445 and 459 bars respectively, so that eqn. (39) becomes

$$\tau = 445 - 459 \cdot \cos 2(30° - \beta).$$

This equation gives good agreement with the various values of cohesive strength estimated from the experimental data. The fact that the equation gives rise to a small negative cohesive strength rather than a small positive cohesive strength for $\beta = 30°$, is due to the statistical approach and the spread of experimental results.

It should be noted, that in the derivation of eqn. (41) it was assumed that a and β were between $0°$ and $90°$. However, if it is now assumed that a is negative, $\tau_m \cdot \sin 2a$ in eqn. (41) must be replaced by $|\tau_m \cdot \sin 2a|$. Then, since $\cos 2a = \cos(-2a)$ while $\sin 2a = -\sin(-2a)$, it may be inferred that the left-hand side of eqn. (41) has greater values for $a > 0$ than for $a < 0$. Thus, the conditions of failure are attained more readily for positive values of a and therefore failure is likely to take place along only one plane instead of along the two possible conjugate shears which may develop in isotropic material.

Moreover, the experimental and theoretical results indicate that rock with a well-developed anisotropy may fail in a brittle manner along a shear plane which forms an angle of greater than $45°$ to the direction of maximum principal stress. Hence, it would be incorrect to assume that the existence of such large angles of shear necessarily indicates that failure has followed plastic deformation.

With reference to the angles of fracture for normal faults and thrusts determined by Sax: these structures occur in sediments where the bedding represents a planar anisotropy. In general, the normal faults cut the bedding at high angles and are therefore largely unaffected by the anisotropy. The thrusts, however, tend to form at low angles to the planes of anisotropy and the angle of fracture is consequently reduced.

FURTHER CONDITIONS OF STRIKE-SLIP FAULTING

A condition approximating to the standard state may possibly develop in sediments which have undergone consolidation and compaction under gravitational loading. However, sediments which have become cemented soon after their deposition, or which were compacted in a previous phase of subsidence, can be dealt with as elastic bodies. Price (1959) used this approach to analyse the conditions of stress which must be satisfied before wrench faults may develop in such rocks. This analysis is of importance when discussing the conditions which can give rise to jointing, so it will be given below in some detail.

The basic equations relating triaxial stress and strain, which can be derived from eqn. (10), are given below

$$\varepsilon_x = \frac{1}{E}\left[\sigma_x - \frac{1}{m}\left(\sigma_y + \sigma_z\right)\right]$$
$$\varepsilon_y = \frac{1}{E}\left[\sigma_y - \frac{1}{m}(\sigma_x + \sigma_z)\right] \quad (44)$$
$$\varepsilon_z = \frac{1}{E}\left[\sigma_z - \frac{1}{m}(\sigma_x + \sigma_y)\right]$$

where $\varepsilon_{x,y,z}$ are the strains in mutually perpendicular directions; and $\sigma_{x,y,z}$ are the corresponding stresses. E is Young's modulus and m is Poisson's number, the reciprocal of Poisson's ratio.

These equations can be used to establish the stress conditions in the earth's crust. Consider a small unit cube of rock in the crust subject to a vertical gravitational stress σ_z. Lateral expansion is completely restricted by the pressure of the surrounding rock; i.e. $\varepsilon_y = \varepsilon_x = 0$ and, therefore, from eqn. (44), it follows that the lateral pressure is given by

$$\sigma_y = \sigma_x = \sigma_z/(m - 1). \quad (45)$$

The value of m varies somewhat from rock type to rock type. It is also a function of pressure, for Phillips has shown that in uniaxial compression the value of m decreases as load increases. Price (1958) has established a similar relationship in conditions of triaxial compressive stress. The value of *m* and the manner in which horizontal pressure is related to depth of burial (assuming pore-water pressure is zero) for two moderately competent coal measure sandstones, each with a uniaxial, short-term, air-dry compressive strength in excess of 20,000 lb/in², are shown in Fig. 29.

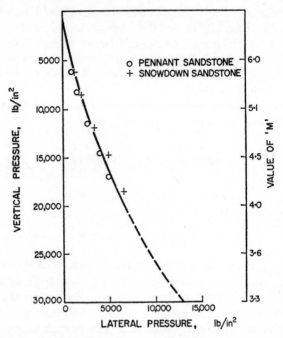

FIG. 29. Relationship between vertical and horizontal pressure in typical competent sandstones, also showing values of Poisson's number at the various stress levels.

Consider now the effect of a horizontal stress c_y superimposed upon the stress field due to gravity. Then, neglecting the effect of the very small increase of overburden which results from the compressive stress c_y, the new stresses in the y- and x-directions are given by

$$\sigma'_y = \frac{1}{m-1} \cdot \sigma_z + c_y \tag{46}$$

and

$$\sigma'_x = \frac{1}{m-1} \cdot \sigma_z + \frac{1}{m} \cdot c_y. \tag{47}$$

Clearly, σ_z, σ'_y and σ'_x are principal stresses and, when their relative values are known, the greatest, intermediate and least principal stresses can be designated σ_1, σ_2 and σ_3 respectively.

According to the Navier–Coulomb criterion of failure, two conditions must be satisfied before wrench faults can develop:

(1) The intermediate principal stress must act vertically at the instant of failure.

(2) The shear stresses must be sufficiently large to cause failure of the rock.

From eqns. (45), (46) and (47) it follows that if the load due to gravity is to represent the intermediate principal stress (i.e. $\sigma_z = \sigma_2$) then the value of the compressive stress c_y must fall within the limits represented by

$$\sigma_z \cdot \frac{(m-2)}{(m-1)} \leqslant c_y \leqslant \sigma_z \cdot \frac{m(m-2)}{(m-1)} . \tag{48}$$

Obviously, the limiting conditions of c_y are determined by the value of m. Thus, if $m = 4$, in order that the intermediate principal stress may act vertically, c_y must fall between the limiting values $0.67\,\sigma_z$ and $2.6\,\sigma_z$. By substituting these values in eqns. (46) and (47), it can be shown that the ratio of the greatest to least principal stress σ_1/σ_3 cannot exceed 3.0. Indeed, it can readily be shown from eqns. (46), (47) and (48) that the maximum ratio of σ_1/σ_3 is given by

$$\sigma_1/\sigma_3 = (m-1). \tag{49}$$

Consider now the second condition which must be satisfied before wrench faults may develop, namely, that the ratio σ_1/σ_3 is sufficiently large to cause shear failure.

Assume that the two coal measure sandstones, the experimental data of which are given in Fig. 29, fail under long-term stress according to the relationship

$$\sigma_1 = \sigma_l + K \cdot \sigma_3.$$

It has been shown that the value of K for these rock types is greater than 4·0, while the short-term uniaxial strength is greater than 20,000 lb/in². If it is assumed that the long-term fundamental strength σ_l of these rock types is approximately 50 per cent of their short-term uniaxial strength and that K remains unchanged (see Chapter 1), then the relationship between principal stresses at failure will probably approximate to

$$\sigma_1 = 10,000 + 4 \cdot 0 \,.\, \sigma_3 \text{ lb/in}^2.$$

Since a limit of 20,000 ft has been set on the zone of brittle failure, the maximum value which the least principal stress can obtain is approximately 20,000 lb/in². The ratios between principal stresses at failure for values of σ_3 up to 20,000 lb/in² are given in Table 6.

TABLE 6

σ_3 lb/in²	σ_1 lb/in²	σ_1/σ_3
0	10,000	∞
5000	30,000	6·0
10,000	50,000	5·0
15,000	70,000	4·66
20,000	90,000	4·5

However, it will be seen from Fig. 29 that for a value of $\sigma_3 = 5000$ lb/in², σ_1 is approximately equal to 18,000 lb/in² and $m = 4 \cdot 0$. The nearest equivalent stress system in Table 6 is given in line two, where $\sigma_1/\sigma_3 = 6 \cdot 0$. Thus, for shear failure to take place when the least principal stress has a value of 5000 lb/in² the ratio of principal stresses must be 6·0. However, it ·has been shown that for the postulated conditions of compression, the maximum ratio of σ_1/σ_3 is given by $(m - 1)$, so that if $\sigma_3 = 5000$ lb/in², $m = 4 \cdot 0$ and the maximum ratio of the greatest and least principal stresses is 3·0. This is smaller than any of the ratios cited in Table 6 necessary to cause shear failure. Consequently, it can be concluded that wrench faults will not develop in the conditions postulated. One is forced to the conclusion that, even for those rocks which may be considered as elastic and brittle from the earliest times, wrench faults can only develop if there is a lateral expansion concomitant with a lateral compression at right angles, which has the effect of increasing the ratio of the greatest to least principal stresses.

STRUCTURES ASSOCIATED WITH STRIKE-SLIP

Clearly, in districts where two sets of contemporaneous wrench faults intersect at an acute angle, and the displacement along the faults is such that the acute wedge of rock moves towards the line of intersection of the fault planes, it is plausible to interpret the structures as a system of complementary shears. However, if the sense of movement on both sets is the same (i.e. both left-handed or both right-handed) they obviously cannot be interpreted as complementary fractures. Such structures are represented in Fig. 30,

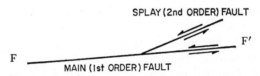

SPLAY (2nd ORDER) FAULT

F

F'

MAIN (Ist ORDER) FAULT

FIG. 30. Main or first-order fault with off-shoot, splay or second-order fault.

which shows a main fault plane *FF'* and an off-shoot or *splay* fault which makes a small angle with it. When interpreting such structures, it may be suggested that one fault is older than the other and is the result of a different stress system, or else the fracture direction of one fault has been influenced by pre-existing planes of weakness. Indeed, in some instances these interpretations are almost certainly correct. But it is probable that in many cases the splay fault (and other structures which will be considered later) are genetically related to the main fault.

Differential movement along a fault plane will bring about an alteration in the stress system which initiated the faulting. The greatest principal stress is reduced in value by the movement while the least principal stress is increased. Hence, it will be inferred that a normal fault will cause an increase in the horizontal stress at right angles to the strike of the fault. Correspondingly, a thrust will diminish the horizontal stress.

The actual changes in stresses caused by such faults have not been studied. However, an analysis of the stress changes brought about by movement along a wrench fault has been presented by E. M. Anderson. In his analysis, he assumed that a stress system which is composed of a compressive stress σ and a concomitant tensile stress at right angles of $-\sigma$ is superimposed upon the standard state. It is then supposed that a vertical wrench fault, which is assumed to be an elliptical opening of great eccentricity, is formed and that it bisects the angle between the principal stresses. Using the type of approach adopted by Griffith and Inglis, Anderson then calculated the values and orientation of the axes of principal stress in the vicinity of the fault plane after movement had taken place.

He found that near the central portion of the fault, the stresses causing fault movement were relieved. However, the stress intensity increased at right angles to the fault plane, so that at a distance of 0·4 times the length of the fault the original level of stress is equalled, or exceeded. Hence, at this distance, it is possible that a second main fault may develop. Anderson found that there is also an increase in the intensity of the stresses as the ends of the fault plane are approached. Moreover, in the vicinity of the fault ends the axes of principal stress swing round and tend to become approximately parallel and normal to the fault plane. It is proposed by Anderson that the orientation of the principal stresses and the high stress intensities which tend to develop near the fault ends may account for the mode of occurrence of splay faulting.

In the field, however, wrench faults are not open fractures. Consequently, an analysis of the type presented by McClintock and Walsh, which takes into account the frictional forces along the fault plane would be more apposite.

The reorientation of stresses along wrench faults due solely to such frictional drag has been dealt with by McKinstry. The analysis is intended to apply only to steeply dipping wrench faults, for in such structures, the forces due to gravity act in a direction which is

approximately parallel to the fault plane and consequently has no
appreciable component acting either normal or parallel to the fault.

For ease of presentation of his thesis, McKinstry assumed that
strike-slip movement was in progress on two parallel planes as
indicated in Fig. 31a and then considered the stresses acting in the
slab of rock enclosed between the two fault planes.

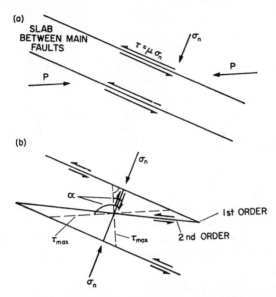

FIG. 31. (a) Normal and shear stress acting on slab between two
parallel vertical wrench faults. (b) Planes of maximum shearing stress and
planes of second-order shear failure in the fault slab (after McKinstry).

As indicated, there will be a stress σ_n normal to the fault planes
and shear stress $\tau = \mu_s \cdot \sigma_n$ acting along the fault planes. Under these
stress conditions it can be shown (see Salmon, p. 32) that the planes
of maximum shearing stress will be determined by the equation

$$\tan 2a = \frac{\sigma_n}{2\tau} = \frac{\sigma_n}{2\sigma_n \cdot \mu} = \frac{1}{2\mu_s} \tag{50}$$

where a is the angle between the planes of maximum shearing stress
and the normal to the fault (see Fig. 31b).

If the angle of friction is used in eqn. (50), then

$$\tan 2a = 1/(2 \tan \varphi_s). \tag{51}$$

McKinstry then argues that, as a rough approximation, $\tan 2\varphi$ can be taken as equivalent to $2 \tan \varphi$, so that

$$\tan 2a = \frac{1}{2 \tan \varphi_s} = \cot 2\varphi_s = \tan (90° - 2\varphi_s)$$

$$2a = (90° - 2\varphi_s) \text{ or } (270° - 2\varphi_s)$$

and $\qquad a = (45° - \varphi_s) \text{ or } (135° - \varphi_s).$

This gives the angle between the planes of maximum shearing stress and the fault plane. If it is assumed that the angle of internal friction φ_1 and the angle of sliding friction are equal, then second-order shear fractures would make an angle of $\frac{1}{2} \varphi_i$ to the planes of maximum shearing stress. Hence, the angle which such shear fractures would make with a normal to the primary, or first-order fault planes would be $45° - 3\varphi/2$ and $135° - \varphi/2$ respectively. Thus, if $\varphi = 30°$, these shears would make an angle of $0°$ and $120°$ to a normal to the primary fault planes, i.e. the second-order shears would make an angle of $90°$ and $30°$ with the first-order shears.

McKinstry points out that the special case of a slab of rock situated between two parallel faults has been used purely for ease of representation of the argument: it is suggested that the same analysis and conclusions apply to the stress orientations adjacent to a single fault plane.†

He has further pointed out that the second-order shears could, in turn, give rise to third-order shears, but suggests that the available energy may be rapidly dissipated by movement along the first-order fault planes so that lower order shears may fail to develop or will be of minor importance.

Moody and Hill, on the other hand, consider that the development of stress concentrations near the end of faults (demonstrated

---

† It should be noted that McKinstry does not take into consideration the influence of stress parallel to the fault zone. Nevertheless, despite the inadequacies of this and the previous analysis, there is recent experimental evidence [Colback – private communication] which demonstrates that secondary fractures do in fact develop in the position and possess the general orientations suggested by the theories of Anderson and McKinstry.

by Anderson) coupled with the mechanism suggested by McKinstry
can in fact give rise to the development of important second- and

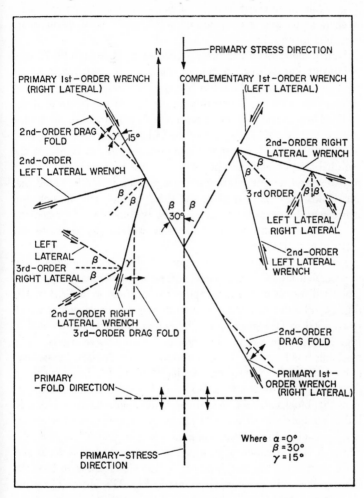

FIG. 32. Diagram of first-, second- and third-order faults and associated fold directions which may theoretically result from a north–south primary compression (after Moody and Hill).

third-order shears. Indeed, they develop the hypothesis that these orders of wrench faults may be accompanied by folds and thrusts. The idealized interrelationship of first-, second- and third-order wrench faults and the associated folds which may develop are indicated in Fig. 32.

Similar concepts have been expressed by Lensen (1959) who points out that the displacement of conjugate, intersecting, contemporaneous wrench faults gives rise to local compressions and tensions in the rock in the vicinity of the line of intersection of the faults. Thus, in the two acute sectors the displacement of the wedge towards the axis of intersection of the faults causes a local shortening of the material. The resultant compressive stresses reinforce the regional stresses and may possibly give rise to secondary folding, thrusting and reverse faulting in these sectors. It will be noted that such folds may be normal to the trend of the main folding and hence may be regarded as cross-folds. By analogy, the extension of the material in the obtuse sectors will locally decrease the regional compression and may give rise to normal faults and dykes.

Lensen (1958) has further suggested that in areas where two non-parallel faults occur, lateral strike-slip movement along these faults will result in the development of a *Horst* or a *Graben*. His thesis can be demonstrated by considering the displacement which is assumed to take place along the two fault planes represented in Fig. 33 in response to the indicated compressive stress. The mechanism is based on the assumption that the frictional resistance to movement along the normal fault (60° dip) is less than that along the plane of the reverse fault (85° dip). Consequently, Lensen suggests that blocks A and C tend to move together relative to block B. If the fault blocks are displaced by a distance x, two lines of reference will be displaced from position a' and b' to a'' and b'' respectively. Thus, there is a tendency for a gap to form between blocks B and C. In such circumstances, block C will subside relative to block B and a graben will develop, as shown in Fig. 33b and c. The same mechanism can be applied to show that lateral movement of the opposite sense can give rise to elevation in block C, resulting in a horst. It may be inferred from Fig. 33 that the amount of subsidence

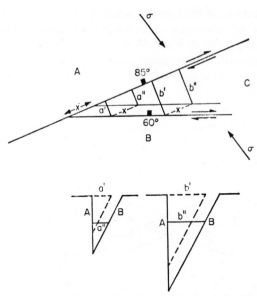

FIG. 33. (a) Plan showing movement along reverse fault under action of
compressive stress σ. (b) Sections through faults demonstrating development of Graben structure (after Lensen).

or uplift of the block is related to the degree of lateral displacement
and the inclination of the line of intersection of the two faults.
Clearly, when the faults become parallel, there will be no vertical
component. Consequently, as Lensen points out, a transcurrent
fault which bifurcates and joins again results in a tilted graben linked
by a stable block to a tilted horst. Lensen quotes examples of fault
patterns in New Zealand where this mechanism is thought to have
operated.

OBLIQUE-SLIP FAULTS

In many localities in the field, striations and slickenside grooving
on fault planes, which indicate the direction of shear movement along
the fault, are neither parallel to the dip nor to the strike of the fault
plane. For example, many such oblique-slip faults occur in the
Girvan district of Scotland (Williams), and may result from one or

more of a number of different mechanisms. Thus, dip or strike-slip faults would be transformed into oblique-slip faults if, subsequent to faulting, the region in which the structures developed underwent tilting. In general, it is suggested that such a mechanism is only likely to give rise to minor deviations from the vertical or horizontal slip directions.

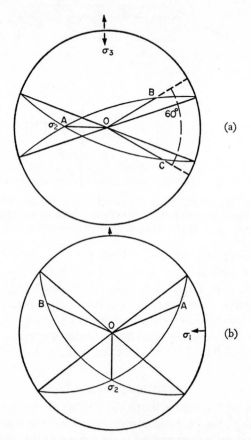

FIG. 34. Stereograms of oblique-slip faults showing direction of slip when (a) σ_3 acts horizontally σ_2 dips 45°W. and σ_1 dips 45°E., and (b) σ_1 acts horizontally σ_2 dips 45°S. and σ_3 dips 45°N.

Oblique-slip faults may develop in a stress field in which the axes of principal stress are inclined to the vertical and horizontal. They may also be a manifestation of planar anisotropy in the rock, when such planes are oblique to all three axes of principal stress.

The development of oblique-slip faults in isotropic material in an inclined stress field is discussed by Williams. If complementary shears develop in such circumstances, the line of intersection of the faults will be parallel to the axis of intermediate principal stress. The direction of shear movement along the fault will be in the plane containing the axes of greatest and least principal stress, while the angle of shear in the $\sigma_1\sigma_3$ plane will be given by $2\theta = 90° - \varphi$.

The faults which develop when the axes of stress are inclined will be hybrid forms, and may be termed "normal-wrench" faults and "reverse-wrench" faults. Such faults are represented stereographically in Fig. 34a and b. The axis of least principal stress represented in Fig. 34a is horizontal; the axes of intermediate and greatest principal stresses are inclined at 45° to the west and east respectively. If φ is assumed to be 30°, $2\theta = 60°$ and, since shear movement will take place in the $\sigma_1\sigma_3$ plane, the lineations or slickensiding which will result from such failure are therefore represented by OB and OC. Such faults are normal-wrench faults.

Reverse-wrench faults are represented in Fig. 34b. In this example, the axis of greatest principal stress is horizontal while the axes of the intermediate and least principal stresses are inclined at 45° to the south and north respectively. The direction of slip along these fault planes is represented by the lines OA and OB.

As an alternative to stereographic methods of analysis and representation, Williams presents a series of equations (not given here) which permit the various essential parameters of the shear planes to be calculated.

Analyses of situations which can give rise to stress systems in which axes of principal stress are inclined to the horizontal and vertical are discussed later.

Oblique-slip faulting may also develop as a result of failure along a plane of anisotropy, or by a regeneration of movement along a pre-existing fault plane.

Bott has shown that the direction of maximum shearing stress in any plane XYZ, (see Fig. 35), which is at an angle to the three axes of principal stress is given by

$$\tan \theta = \frac{n_*}{l_* m_*} \left[m^2_* - (1 - n^2_*)\frac{\sigma_z - \sigma_x}{\sigma_y - \sigma_z} \right] \qquad (52)$$

where θ is the pitch of the maximum shearing stress (i.e. the angle within the XYZ plane between the strike direction and the direction of the maximum shearing stress); σ_x, σ_y and σ_z are the principal stresses and l_*, m_* and n_* are the direction cosines of the plane.

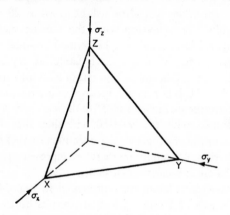

FIG. 35.   Plane of weakness orientated obliquely to all three axes of principal stress.

If it is assumed that the XYZ plane represents any plane of weakness, and that the stress conditions are also sufficient to cause shear, then movement along the plane will be in the direction of maximum shearing stress as defined by eqn. (52). Hence, depending upon the orientation of the stresses with respect to a specific plane, and upon the relative values of the principal stresses one to another, then oblique-slip faulting may occur in any possible direction in that plane.

However, it may be inferred from eqn. (52) that the reverse process of determining the axis of stress from the angle θ is not possible without other data. In other words, a unique determination of the orientation of the axes of principal stress from data and measurements relating to the direction of slip taken from fault planes in the field is often extremely difficult, or even impossible.

OVERTHRUST FAULTING AND PORE-WATER PRESSURE

Overthrusts are faults which need special mention, for they are structures which do not conform to the simple classification of faults dealt with in a previous section. Overthrusts form extensive planes along which large masses of rock have been displaced over distances which are usually measurable in miles, or even tens of miles. A fundamental characteristic of these structures is that movement has taken place along a plane which initially dipped at 10° or less; although, of course, the orientation of the thrust plane may subsequently be altered by folding.

Overthrusts have been recognized in many areas throughout the world. For example, there is the Moine overthrust, in the metamorphic rocks of Scotland, with a horizontal displacement of not less than 10 miles (16 km) (Peach and Horne). The great thrust in the Scandinavian peninsula has a horizontal displacement of more than 80 miles (Tornebohm). The major overthrust in the Glarus area of the Swiss Alps exhibits a displacement of approximately 25 miles (Oberholzer). Large overthrusts have been reported in the Himalaya of India (West) and in the Peruvian Andes (Baldry, Brown). Many such structures are to be found in North America. There are the Taconic Thrust of the Appalachians (Cady), the Muddy Mt. and Robert Mt. overthrusts of Nevada (Longwell, R. Anderson, Gillully) and the overthrust belt of western Wyoming (Rubey), and along the eastern margin of the Canadian Rockies (McConnel and Willis), to quote but a few of the possible examples.

The existence of such large-scale overthrusts presented a mechanical paradox for many years. However, a solution to the problem was presented in two important and monumental companion papers by Hubbert and Rubey.

Reduced to its simplest possible elements, an overthrust may be approximately represented as a rectangular block sliding along a planar surface AB, which simulates the thrust plane itself. Such a block may be moved in one of three ways: (1) a pressure may be applied to one end of the block; (2) it may move under the influence of its own body force (i.e. it may slide down an inclined plane); or (3) the block may move as a result of both these mechanisms (i.e. it may be pushed down an inclined plane).

The difficulty of applying any of these mechanisms to the problem of overthrusting becomes evident when one considers the dimensions of the rock masses involved in the movement, and the mechanical properties they are likely to possess.

Smoluchowski considered a block of length b, width a and thickness c. Then, if w is the weight per unit volume of the rock, the force F required to move the block is given by

$$F = (a \cdot b \cdot c) \, w \cdot \mu_s$$

where μ_s is the coefficient of sliding friction. Smoluchowski took the block-length to be 100 miles and assumed $\mu_s = 0.15$ (the coefficient of friction of iron sliding on iron). To support this applied stress the rock would need to be able to support the load exerted by a 15-mile-high column of rock without crumbling at the base. However, as we have seen, a more realistic value for the coefficient of friction of competent rock would be 0·5 to 0·6. Hence, the rock column would need to be some 50–60 miles high, i.e. it would have to possess a uniaxial strength of 250,000–300,000 lb/in^2. This figure is 5–10 times greater than the strength of the more competent rocks as measured in the laboratory.

Hubbert and Rubey adopted an alternative approach. They assumed that the maximum stresses which could act on the free face AB are as represented in Fig. 36a. At the upper surface, at point B the maximum horizontal stress is equal to the uniaxial compressive strength of the rock. Lower down the AB face, the bodyweight acts effectively as a confining pressure, and if it is assumed that the relationship between principal stresses at failure is given by

$$\sigma_1 = \sigma_o + K \sigma_3$$

then the maximum stresses which may develop at the AB surface

Plate 3.   Quartz-filled tension joints cut and slightly displaced by a dolomite-filled shear joint.

without the block failing, will be distributed as indicated in Fig. 36b. Under such a stress, they showed that the maximum length of the block which will slide is given by

$$y_{max} = \frac{\sigma_o}{\rho \cdot g \cdot} \tan \varphi + \frac{K \cdot \zeta}{2 \tan \varphi} \qquad (53)$$

where ζ is the thickness of the block.

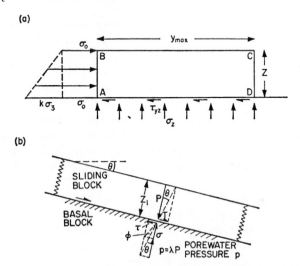

FIG. 36.   (a) Block resting on a plane simulating an overthrust acted on by stress system (after Hubbert and Rubey). (b) Gravitational sliding of sub-aerial block with pore-water pressure acting at the "glide plane" (after Hubbert and Rubey).

They point out that if the block were longer than y_{max}, and the stresses exceed those indicated in Fig. 36a, the block as a whole will not move, but instead will fail at the AB end as an ordinary thrust.

There is, however, a limit to the analysis which has not been pointed out by Hubbert and Rubey, and this relates to the maximum pressure which can develop in the vertical direction at the other free vertical face CD. Clearly, the maximum vertical pressure which can exist at point D is equal to the uniaxial compressive strength of the rock. In their analysis, it was assumed that the uniaxial strength is

7×10^8 dyn/cm^2, or approximately 10,000 lb/in^2. Since ρ is taken to be 2·31, the maximum thickness which the block can attain without beginning to crumble at the edge of the base is approximately 10,000 ft, or very approximately 2 miles. This limitation consequently invalidates some of the data presented by Hubbert and Rubey. It does not, however, alter the conclusions which may be drawn from their analysis.

If φ is taken as equal to 30°, then since, from eqn. (20), $K = (1 + \sin \varphi)/(1 - \sin \varphi)$, it follows that $K = 3\cdot0$. If one now takes the further limitation that the maximum value of ζ is 2 miles, then the maximum length of the block which will slide under the postulated stress conditions is 5 miles. If one wishes to consider the conditions of sliding of a thicker block it is necessary to postulate that the block consists of correspondingly stronger material. Clearly, with the mechanical conditions and physical properties postulated, it is impossible to move a block of the length which field observations demand (i.e. 80–100 miles) along a horizontal plane.

It has long been realized that gravitational sliding is a mechanism which is unaffected by the limitations set by the strength of the rocks. However, this mechanism is beset by equally stringent limitations, for, as may easily be shown, the least angle φ of slope which must exist before sliding will take place is determined by the coefficient of sliding friction. By resolving the components of the body weight parallel and normal to the slope, it can be shown that sliding will take place when

$$\mu_s = \tan \varphi. \tag{54}$$

Reasonable values for μ_s are from 0·5 to 0·6; for such values φ will be close to 30°.

A slope of 30° which extends in the dip direction for 100 miles is clearly out of the question.

The position is only slightly improved if both mechanisms are invoked and a block is pushed down an inclined plane.

It may be inferred that if either mechanism is to be used to explain overthrust faulting, it is necessary, in some way or other, to reduce the effective frictional resistance to the sliding of the block. Now it has been tacitly assumed that the block and sliding surface have been dry and unlubricated. There is probably little likelihood of reducing, in the field, the coefficient of sliding friction by lubrication. However, it is possible to reduce the effective normal stress acting on the thrust plane, and hence, to reduce the resistance to sliding, by introducing into the problem the effect of pore-water.

The mechanical effects of pore-water have been studied and demonstrated in the laboratory on soils (Terzaghi), concrete (McHenry), and rock (Handin *et al.* 1963). These effects may be deduced in the following way.

Consider a cylindrical specimen of rock which has been impregnated, so that the whole of the pore-spaces are filled with water, and then sealed in a flexible, watertight jacket. If such a specimen

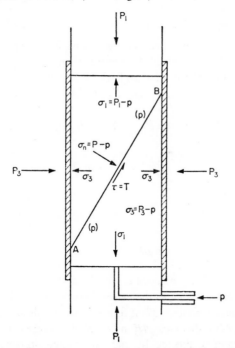

Fig. 37. Active stresses and pore-water pressures which develop in a saturated specimen subjected to triaxial compression (after Hubbert and Rubey).

is deformed in a triaxial compression testing machine under a confining pressure P_3, which acts on the curved surfaces of the cylinder, and an axial pressure P_1, it follows that the pore spaces will be deformed. Since the pores are filled with water which cannot escape and which can be regarded as virtually incompressible, a hydraulic pressure p builds up in the pore spaces and is known as the pore-water pressure. In some tests, especially if the rock is coarse-grained and has a large porosity, the value of p can be controlled as indicated in Fig. 37. It will be seen from this figure that the effective stresses σ_1 and σ_3 (i.e. the principal stresses which may bring about failure) are given by

$$\sigma_1 = P_1 - p$$

and

$$\sigma_3 = P_3 - p. \tag{55}$$

Now the normal and shear stresses acting on a plane AB inclined at an angle θ to the axis of the specimen are given by eqns. (6) and (7). Expressing the normal and the shear stresses in terms of the external applied stress P_1 and P_3 and the pore-water pressure p, one obtains

$$\sigma_n = \frac{(P_1 - p) + (P_3 - p)}{2} - \frac{(P_1 - p) - (P_3 - p)}{2} \cdot \cos 2\theta$$

$$\tau = \frac{(P_1 - p) - (P_3 - p)}{2} \cdot \sin 2\theta$$

which reduces to

$$\sigma_n = \frac{P_1 + P_3}{2} - \frac{P_1 - P_3}{2} \cdot \cos 2\theta - p = P_n - p \tag{56}$$

$$\tau = \frac{P_1 - P_3}{2} \cdot \sin 2\theta = T \tag{57}$$

where P_n and T are the normal and shear stresses on the AB plane which are entirely due to the external stresses P_1 and P_3.

From eqns. (56) and (57) it follows that the shear stress which is generated along the plane AB in Fig. 37 due to the external stresses, is completely unaffected by the pore-water pressure, while the effective stress normal to the plane is that generated by the external stresses P_1 and P_3 minus the pore-water pressure p.

The validity of these conclusions can very readily be demonstrated by conducting triaxial compression experiments using unjacketed specimens. If the rate of deformation is sufficiently slow, so that migration of the pore fluid is able to take place, then the confining pressure P_3 and the pore-water pressure p are identical. It may be inferred from eqns. (56) and (57) that in such tests, failure will take place when $P_1 - P_3$ is a constant equal to the uniaxial wet strength of the rock. It has been shown by Handin and Hager and others that this relationship is in fact obtained for a very considerable range of confining pressures. Hence the effects predicted by the simple analysis given above adequately represents the mechanical influence of pore-water pressure.

In the upper layers of the earth's crust the rocks will be porous and possibly fractured, and in most areas of the world, below a depth of a few tens of feet, these void spaces will be filled with water or very occasionally with oil or gas.

If it is assumed that the void spaces in extensive rock masses are, in general, inter-connected, then the pore-water pressure is likely to be related to the depth of burial by the expression

$$p = \rho_w \cdot g \cdot \zeta \tag{58}$$

where ρ_w is the density of water. This will give rise to a *normal* or *hydrostatic* pressure which increases linearly with depth.

The pore-water pressure encountered in deep bore-holes is frequently greater than the hydrostatic pressure and, indeed, has been known to approach the *geostatic* or *lithostatic* pressure which is given by

$$P_G = \rho_b \cdot g \cdot \zeta \tag{59}$$

where ρ_b is the bulk density of the water-saturated rock. It is convenient to relate the ratio of the pore-water pressure and the geostatic pressure by the parameter λ such that

$$p = \lambda P_G \tag{60}$$

where λ will have limiting values of 0 and 1·0.

Hubbert and Rubey quote a large number of examples from Pakistan, Iran, the Gulf region of Texas and elsewhere, where sediments occur with values of λ which range from 0·8 to 0·95.

Such anomalous pressures, which may be restricted to a single layer, or lens, are best developed in clay sediments which are readily compressible, of relatively large porosity and low permeability. It is suggested that the abnormal pressures may develop as a result of progressive loading during sedimentation, or by tectonic compression.

These abnormally high pore-water pressures are of the greatest significance when considering the overthrust problem. Consider the sliding of a sub-aerial block down an inclined plane, where, it is assumed, the pore-water pressure at the level of the thrust plane is

$$p = \lambda P \tag{61}$$

where $P = \rho_b \cdot g \cdot \zeta \cdot \cos \theta$ (see Fig. 36b). Similarly, the tangential stress T is given by

$$T = \rho_b \cdot g \cdot \zeta \cdot \sin \theta. \tag{62}$$

Clearly

$$T/P = \tan \theta. \tag{63}$$

From eqns. (56), (57), (61) and (62) the corresponding components of effective stress σ and τ are given by

$$\sigma = P - p = (1 - \lambda)P$$

and

$$\tau = T.$$

As previously noted, the conditions for sliding are

$$\tau/\sigma = \tan \varphi \quad \text{or} \quad \tau = \sigma \tan \varphi$$
$$= (1 - \lambda)P \cdot \tan \varphi.$$

From eqn. (63), $T = \tau = P \cdot \tan \theta$.

Therefore

$$(1 - \lambda) P \cdot \tan \varphi = P \cdot \tan \theta$$

or

$$\tan \theta = (1 - \lambda) \tan \varphi. \tag{64}$$

The angle of slope θ down which a block will slide is given in Table 7 as a function of λ where, as previously, it is assumed that $\varphi = 30°$.

Hubbert and Rubey presented an analysis of the maximum length of block which can be pushed down an inclined plane across which the pore-water pressure $p = \lambda P$. It was shown that,

TABLE 7

λ	$1 - \lambda$	$(1-\lambda)\tan\varphi = \tan\theta$	θ
0	1·0	0·577	30·0°
0·4	0·6	0·346	19·1°
0·8	0·2	0·115	6·6°
0·9	0·1	0·058	3·3°
0·95	0·05	0·029	1·6°

for the type of stress distribution on the end-face previously described, the maximum length of block y_{max} is given by

$$y_{max} = \frac{1}{(1-\lambda)\tan\varphi - \tan\theta}\left[\frac{\sigma_0}{\rho_b \cdot g \cdot \cos\theta} + \frac{K - (1-K)\lambda \cdot \zeta}{2}\right].$$

(65)

As in the earlier example, the value of ζ is limited by the strength of the rock forming the block. For small values of θ and strength 10,000 lb/in² used in the previous example, the maximum value for ζ is again approximately 2 miles. The values of y_{max} given in Table 3 of Hubbert and Rubey's paper, which are based on an assumed thickness of approximately 4 miles for the same rock strength, are therefore not strictly valid.

The maximum length of block in miles for a thickness of 2 miles, and the various values of θ and λ, are given in Table 8. It will be seen that for higher values of λ this mechanism can readily account for the movement of overthrust blocks comparable in length with those observed in the field.

Obviously, the difference in altitude of the top and bottom of the postulated slope must have certain limits. A difference of 3 miles is probably a reasonable limiting value. This, in turn, sets a limit on the possible length of the thrust block, for

$$y \cdot \sin\theta = 3 \text{ miles.}$$

Thus, for a slope of 1° the maximum length of the block y will be 170 miles. Such an estimated length of thrust block is compatible with the field evidence. Moreover, providing the rock beneath the

TABLE 8

θ \ λ	0·0	0·8	0·9	0·95
0	8	28	53	106
1°	8·5	33	76	268
2°	9	40	140	—
4°	9·5	64	—	—

thrust block possesses a high pore-water pressure, so that λ approaches 0·95, a block of this length composed of rock of quite modest strength is capable of being thrust down a slope of 1° (See note, p. 109).

THRUST DEVELOPMENT IN FOLD BELTS

Orogenic belts which develop in geosynclinal sediments are a common tectonic feature. The folds which develop in such belts are frequently asymmetrical and are often overturned in one direction and are associated with thrusts which fit the same "movement picture". Examples of such fold belts occur in the Scottish Highlands, the Alps, the Rocky Mountains and elsewhere. A section through a portion of such an orogenic belt is represented diagrammatically in Fig. 38a and indicates that the folds and thrusts tend to die out near the margin of the belt and grade into relatively undisturbed strata. It is highly debatable whether one may apply the concept of brittle fracture to the development of faults in such plastically deformed rocks. However, it is of interest to see to what extent an analysis based on elastic theory fits the observed data.

It was at one time customary to attribute such fold belts to a "one-sided" or "active" pressure or force, which acted in the

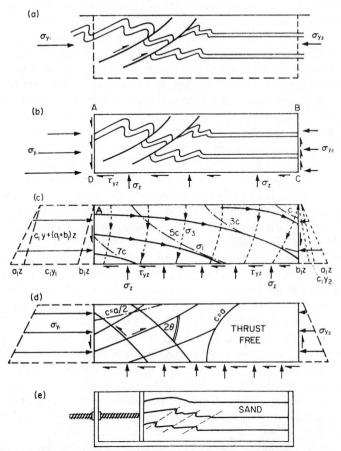

Fig. 38. (a) Diagrammatic section through portion of fold belt with thrusts concave upwards and unfolded "Foreland". (b) Intuitively derived boundary stress system which could give rise to structures represented. (c) Stress trajectories and lines of equal shearing stress which result from the specific stress system indicated. (d) Conjugate shears which may develop from the same specific boundary stress conditions and showing area which may be free from thrusts. (e) Diagrammatic representation of sand box experiments with boundary conditions similar to those indicated in (b), (c) and (d) showing type of deformation shown in (a). (a), (b) and (e) after Hubbert. (c) and (d) after Hafner.

general direction of the movement. However, the rate of tectonic deformation is usually exceedingly slow so that in such geological problems, one is dealing with a static state of affairs and, hence, the forces acting at any one time must be in equilibrium.

Now it is permissible to choose any arbitrary and imaginary surface inside a stressed medium as a boundary surface, without causing a disturbance in the internal stress distribution, provided a system of stresses which are equivalent to the internal stresses is applied to the boundary surfaces. Hence, one may select the rectangle $ABCD$ as representing the boundary surfaces to the region being studied. For the sake of simplicity, it is assumed here that the problem is two-dimensional. Since the problem is a static one, the boundary stresses must be in complete equilibrium and, consequently, the stress system indicated in Fig. 38a must be considerably modified. Thus, there will be a vertical stress σ_z acting along the CD boundary which is equal to the gravitational stress.

Since the strata to the left of the block are more intensely deformed than those to the right, the inference that σ_{y_1} is much greater than σ_{y_2} is inescapable. Consequently, to bring the system into equilibrium in the horizontal direction, it is necessary to postulate a horizontal shearing stress τ_{yz} acting along the base of the block. Since a shearing stress cannot exist along the free surface of the block AB, one must postulate shearing stresses acting along the boundaries AC and BD which will counter the turning couple of the shearing stress acting along the base of the block. These shear stresses acting along the vertical boundaries will be zero at the surface and equal to the basal shearing stress at the points C and D (see Fig. 38b).

It may be inferred from these boundary stresses that the axes of principal stress within the block are not everywhere vertical and horizontal. However, to enable the precise orientation of the axes of principal stress to be ascertained, a more formal quantitative analysis is necessary.

In such a problem, where one is interested in determining the orientation and possible distribution of faults, it is first necessary to establish the orientation of the stress trajectories and the disposition of the lines of equal shearing stress. The stress trajectories are orthogonal curves, such that the axes of principal stress are everywhere tangent to these curves. Hence, in the simple two-dimensional examples which will be considered, one set of curves represents the orientation of the greatest principal stress, and the second orthogonal set represents the orientation of the least principal stress. From the stress trajectories, it is possible to indicate the orientation which shear fractures will have, if they develop, while from the lines of equal maximum shear stress (which connect all points where this quantity is of equal magnitude) it is possible to indicate areas where faults are most likely to form.

The method most usually adopted in such an analysis is based upon the use of an Airy stress function.

The conditions which such functions must satisfy are represented by a biharmonic equation, the detailed derivation of which is involved; however, the mathematics linking the steps in its derivation, indicated in the following paragraphs, can be obtained by referring to Jaeger (1962).

The biharmonic equation is based in part upon the geometrical fact that when an elastic body is deformed, the infinitesimal shear strain ɣ at any point in the body is related to the increments of strain e_y and e_z in the y- and z-directions by the following *compatibility equation*

$$\frac{\partial^2 e_z}{\partial y^2} + \frac{\partial^2 e_y}{\partial z^2} = \frac{\partial^2 \Upsilon_{zy}}{\partial z \, \partial y}. \tag{66}$$

One of the conditions postulated, when deriving the biharmonic equation, is that of plane strain, i.e. $e_x = 0$, so that $\sigma_x = \dfrac{1}{m} \, (\sigma_z + \sigma_y)$

From this and other stress strain relationships (Jaeger 1962, § 14, p. 60) it is possible to express the compatibility conditions of eqn. (66) in terms of stress as follows

$$\left(\frac{\partial^2}{\partial y^2} + \frac{\partial^2}{\partial z^2} \right) (\sigma_y + \sigma_z) = 0. \tag{67}$$

In addition, geological problems of the type under discussion are essentially static ones, so that the boundary stresses must be in equilibrium.

If it is supposed that the boundary stresses can be represented by a function Φ, then it can be shown (Jaeger 1962, § 34) that the conditions for static equilibrium are automatically satisfied if

$$\sigma_y = \frac{\partial^2 \Phi}{\partial \zeta^2}, \ \sigma_z = \frac{\partial^2 \Phi}{\partial y^2}, \ \tau_{yz} = -\frac{\partial^2 \Phi}{\partial \zeta \partial y} \tag{68}$$

when body forces are absent and by the expressions

$$\sigma_y = \frac{\partial^2 \Phi}{\partial \zeta^2}, \ \sigma_z = \frac{\partial^2 \Phi}{\partial y^2} + \rho \cdot g \cdot \zeta, \ \tau_{yz} = \frac{-\partial^2 \Phi}{\partial \zeta \partial y}. \tag{68a}$$

when gravity represents the only body force.

Since the conditions represented by eqn. (67) and by eqn. (68) must simultaneously be satisfied if the mathematical treatment is to be valid, then these two equations may be combined into the biharmonic equation

$$\frac{\partial^4 \Phi}{\partial y^4} + \frac{2\partial^4 \Phi}{\partial y^2 \ \partial \zeta^2} + \frac{\partial^4 \Phi}{\partial \zeta^4} = 0 \tag{69}$$

which is sometimes abbreviated to

$$\nabla^2 \left(\nabla^2 \ \Phi \right) = 0.$$

In an analysis of problems of the type under discussion, it is necessary to establish an Airy stress function Φ which will satisfy eqn. (69) and also provide boundary conditions which are apposite to the problem in hand.

It will be seen that the stress function Φ is obtained by a solution of a fourth-order differential equation, so that it will contain four independent constants of integration. Since any stress system must satisfy the boundary conditions at the earth's surface ($\sigma_z = 0$ and $\tau_{yz} = 0$, when $\zeta = 0$), two of the constants must be used for this purpose. The remaining two, however, may be used for modification of the boundary stresses.

It follows from the principal of superposition of stresses, that if there are two stress functions Φ_1 and Φ_2, each of which represents a correct solution of eqn. (69), then the sum of the two functions is also a valid solution, i.e.

$$\Phi = \Phi_1 + \Phi_2. \tag{70}$$

It will be remembered that Anderson introduced the idea of the standard state. An important feature of this concept is that the body forces due to gravity can be incorporated into a stress system which is essentially hydrostatic. It therefore contributes nothing to the shearing stresses and, consequently, has no influence upon the configuration of the stress trajectories. If the standard state is represented by stress function Φ_1, it follows from eqn. (70) that one may superimpose upon the standard state any other valid stress system Φ_2 caused by the boundary conditions. Anderson has named systems derived from Φ_2, *supplementary stresses.*

It has been shown by Hafner that the standard state may be derived as follows. He selected the Airy stress function in the form of a third-order polynominal

$$\Phi_1 = a_1 y^3 + a_2 y^2 \zeta + a_3 y \zeta^2 + a_4 \zeta^3 \qquad (71)$$

which satisfies eqn. (69). The stress components are derived from eqn. (67) and (71), and are given by

$$\sigma_y = \frac{\partial^2 \Phi_1}{\partial \zeta^2} = 2a_3 y + 6a_4 \zeta$$

$$\sigma_z = \frac{\partial^2 \Phi_1}{\partial y^2} + \rho \cdot g \cdot \zeta = 6a_1 y + 2a_2 \zeta + \rho \cdot g \cdot \zeta \qquad (72)$$

$$\tau_{yz} = \frac{\partial^2 \Phi_1}{\partial y \, \partial \zeta} = 2a_2 y + 2a_3 \zeta.$$

At the earth's surface $\sigma_z = 0$, so that the constants a_1 and a_2 are zero. Also, if it is assumed that a_3 is zero, while a_4 is put equal to $\rho \cdot g/6$, the stress conditions for the standard state are given by

$$\sigma_y = \sigma_z = \rho \cdot g \cdot \zeta$$

and $\qquad \tau_{yz} = 0.$

It has been shown earlier that the lateral pressure due to gravity in a homogeneous solid, with infinite horizontal extent is given by

$$\sigma_y = \sigma_z/(m - 1).$$

Thus, the horizontal stress due to gravity is only equal to the gravitational load when $m = 2$. Usually, m will be greater than 2·0, so that the standard state will generally be made up of two components, one being the lateral pressure due to gravity, and the other a superimposed horizontal stress which is constant in any one horizontal plane, but increases in depth. Clearly, pore-water pressure will contribute to this stress system.

Hafner also demonstrated the application of this method to the analysis of supplementary stresses. In the relatively simple example considered in the following paragraphs, it is assumed that there is a supplementary horizontal stress; but that there is no vertical component of stress acting across any horizontal plane, other than that associated with the standard state. This assumed condition is expressed mathematically as

$$\sigma_z = \frac{\partial^2 \Phi_2}{\partial y^2} = 0$$

for all values of z.

The form of the supplementary stress function Φ_2 can be obtained by integration of the above expression and is given by

$$\Phi_2 = c \cdot f \cdot (z)y + e \cdot y + b \cdot f_2(z) + d. \tag{73}$$

Now from eqn. (69), it follows that the fourth-order differential of functions f_1 and f_2 must be zero. Consequently, the second differentials must be either linear functions of z, constants or zero.

Two of the possible groups of boundary stresses which satisfy these conditions are listed below:

(I) $df_1(z) = 0$ $\qquad\qquad$ $d^2 f_2(z) = z + d$

$\qquad\qquad \sigma_y = bz + d$

$\qquad\qquad \sigma_z = 0$

$\qquad\qquad \tau_{yz} = 0.$

(II) $df_1(z) = z$ $\qquad\qquad$ $d^2 f_2(z) = 0$ $\qquad\qquad$ (74)

$\qquad\qquad \sigma_y = cy$

$\qquad\qquad \sigma_z = 0$

$\qquad\qquad \tau_{yz} = -c_1 z.$

In these equations, b, c and d are arbitrary constants and may be assigned any value including zero. It follows from eqn. (70) that any linear combination of the boundary conditions listed above also represents a permissible system of boundary conditions.

The simplest example of a supplementary stress system is given by group (I) if $b = 0$. The supplementary horizontal stress is then given by $\sigma_y = d$. For these boundary conditions, the stress trajectories and hence the axes of principal stress are everywhere vertical and horizontal. This is the state of affairs postulated by Anderson and can give rise to thrusts which are inclined at an angle of $45° - \varphi/2$ to the horizontal.

Consider now the combination of boundary conditions represented by groups (I) and (II) and the standard state. Since the contribution of d has already been discussed it can now be equated to zero, so that the total stress components are given by

$$\sigma_y = cy + b\zeta + a\zeta$$
$$\sigma_z = a\zeta \qquad (75)$$
$$\tau_{yz} = c\zeta$$

where $a = \rho \cdot g$. Thus, the term $a \cdot \zeta$ represents the hydrostatic standard state and is the only component of stress acting in the vertical direction.

The stresses in the horizontal direction σ_y, when c and b are not zero, are indicated acting on the boundaries of the block (Fig. 38c).

Once the boundary conditions have been established, the orientation of the stress trajectories at any point within the block may be calculated by using the expression

$$\tan 2\beta = \frac{2\,\tau_{yz}}{\sigma_y - \sigma_z} \qquad (76)$$

where β is the angle the stress trajectories at any one point make with the vertical and horizontal.

It will be seen from Fig. 38c that the trajectories are curved and dip away from AC, where compression is greatest. Since the trajectories are curved, it follows that extensive fault surfaces which develop in response to such a stress system should also be curved as indicated in Fig. 38c. It has, in fact, been demonstrated that thrusts which are concave upwards have developed in fold belts which occur in the foothills region of Alberta and elsewhere.

It has been noted that before one can delimit the general zones in which faults are likely to develop, it is necessary to determine the values of the maximum shearing stress, or of the greatest and least principal stresses. The maximum shearing stress is given by

$$\tau_{max} = \pm \left[\left(\frac{\sigma_z + \sigma_y}{2} \right)^2 + \tau_{yz}^2 \right]^{\frac{1}{2}}, \qquad (77)$$

while the intensities of the greatest and least principal stresses are given by

$$\sigma_1 \text{ or } \sigma_3 = \tfrac{1}{2}(\sigma_z + \sigma_y) \pm \tau_{max} \qquad (78)$$

The lines of equal maximum shearing stress are represented in Fig. 38d in terms of multiples of the constant c. The delimitation of the zones in which faults are likely to develop is, as we have seen in Chapter 1, somewhat problematical. However, Hafner suggests that the boundaries of the stable blocks may be indicated in terms of the ratio of c and a. The limits of the stable portions of the block for values of $c = a$ and $c = a/2$ are indicated in Fig. 38d.

To substantiate these analytical conclusions Hubbert (1951) conducted experiments in which he deformed loose sand, using a simple box, or trough, of the type indicated in Fig. 38e, one end of which could be advanced, thereby simulating the postulated analytical boundary conditions. The results of the experiments showed good agreement with those obtained from the analysis.

FAULTS IN SEDIMENTS DUE TO MOVEMENT IN THE BASEMENT COMPLEX

The faulting and structural deformation in sedimentary members may often result from vertical differential movement in the underlying basement complex. It is reasonable to assume that such vertical differential movement may be flexural in character or, if it is due to the movement of fault blocks, it may be abrupt and step-like.

In an analysis of the faults which may develop in superincumbent sediments as a result of such basal movement, Sanford used a technique which is related to that used by Hafner. However, instead of expressing the boundary conditions entirely in terms of stress, Sanford's analysis is based upon specific vertical displacements which are assumed to take place at the lower boundary of the homogeneous isotropic elastic layer. These two methods can give similar results when applied to some problems. However, it is readily apparent that the method used by Sanford is particularly suitable for the study of the faulting of sediments, due to movement in the basement complex.

The analysis in terms of displacement gives rise to two functions Ψ_1 and Ψ_2 which must satisfy the conditions

$$\nabla^2 (\nabla^2 \psi_1) = 0$$
$$\nabla^2 (\nabla^2 \psi_2)^{\cdot} = 0. \tag{79}$$

The solutions for these equations in terms of the stresses σ_z, σ_y and τ_{yz} and the displacements u and v are lengthy and will not be given here.

Sanford considered three specific types of displacement. In the first example, the lower boundary of an elastic layer is assumed to undergo a sinusoidal vertical displacement, but suffers no horizontal

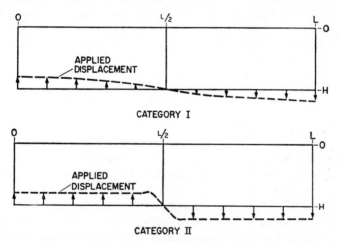

Fig. 39. Types of movement in basement discussed by Sanford.
(a) Sinusoidal uplift. (b) Step-like uplift.

displacement. These conditions represent a gentle flexuring of the basement complex (see Fig. 39a). In the second example, the lower boundary is assumed to undergo a step-like displacement such as would happen during faulting of the basement complex. The form of the analytically applied displacement is shown in Fig. 39b. To obtain this step displacement analytically is a formidable task, for Sanford found it necessary to obtain eight solutions derived from eqn. (79) which were superimposed by means of a Fourier series. As in the first example, it was assumed that there was no horizontal displacement. In the third example, the lower boundary undergoes sinusoidal vertical and horizontal displacement which are 90° out of phase (i.e. as the lower boundary goes down, the adjacent vertical boundary moves inward).

It is emphasized that the vertical movements envisaged in these analyses are very small. In a sedimentary unit some 15,000 ft thick, the maximum movement at the base would be of the order of 30–35 ft. Thus, the movements represented in Fig. 39a and b show a great vertical exaggeration.

It was found that the orientation of the stress trajectories in these examples is influenced only by the dimensions of the layer and, to a small extent, by the value of Poisson's number m.

Sanford also calculated values for the distortional strain energy, using the expression

$$E_d = \frac{1}{12.G} \left\{ \left[\sigma_y - \sigma_z \right]^2 + \left[\sigma_z - \frac{1}{m} (\sigma_y + \sigma_z) \right]^2 + \left[\frac{1}{m} (\sigma_y + \sigma_z) - \sigma_y \right]^2 + 6\tau_{yz}^2 \right\}. \tag{80}$$

The distribution of the intensity of distortional strain energy indicates the regions where fractures are first likely to form, or where material is first likely to yield by plastic deformation.

The stress trajectories and distribution of distortional strain energy for case I are represented in Fig. 40a and b respectively. It will be seen that the zone within 3000 ft of the upper surface of the deformed layer above the anticlinal portion of the basement is in tension and is likely to give rise to a vertical tension crack at point F. There are two regions of high distortional strain energy shown in Fig. 40b. However, shear stresses with a magnitude sufficiently large to cause failure only occur in the region adjacent to the crest of the area of maximum uplift. This derives from the fact that the horizontal stresses in this area are tensile. Since the stress trajectories

are horizontal and vertical in this region, it follow
will form normal faults. The angle between the
and the maximum principal stress will be given $

$$\theta = (45 - \varphi/2).$$

Hence, one may predict that for many rock types
inclined at angles of approximately 60°.

The stress trajectories and distribution of distortional strain
energy for group II are indicated in Fig. 40c and d respectively. It
will be seen that the deformation of the layer is almost completely
restricted to a vertical zone immediately above the "fault line" in
the basement. Since shears form at an acute angle to the axis of
maximum principal stress, it may be inferred that a fracture which
develops in the elastic layer in the vicinity of the basement fault line
will not be vertical. Hence, the fault will run into a zone where the
stress trajectories are curved and, consequently, the fault will also
be curved. In addition, there is a region near the upper surface on the
uplift side of this zone where the horizontal stresses are tensile.
These stresses, as in the previous example, may give rise to vertical
tension gashes or normal faults in the vicinity of F_1.

The results of the third case, in which there is an out-of-phase
lateral movement in addition to the sinusoidal basal displacement,
shows little difference from the conclusions obtained in group I
and, consequently, are not dealt with here.

A feature of the analytical method used in this section is the ease
with which the results may be checked; for displacements of the
type specified in the analysis can readily be applied in scale model
experiments.

The apparatus used by Sanford for this purpose consisted of a
rectangular box, the bottom of which was fitted with three sections
capable of independent vertical movement. Differential movement
of these sections simulated the vertical movement postulated in
group II. The sinusoidal movement condition was simulated by
covering the base with a rubber pad which was displaced by movement in the central section.

The material used in these experiments was sand, which was laid
in successive layers separated by thin marker horizons. In some

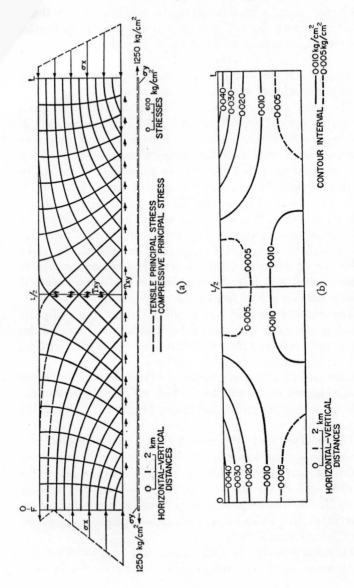

Fig. 40. For caption, see page 105.

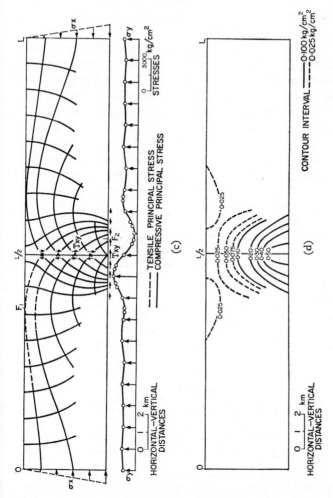

FIG. 40. (a) Stress trajectories in deformed elastic material which result from sinusoidal uplift in basement. (b) Distribution of distortional strain energy resulting from same movement. (c) Stress trajectories in deformed elastic material which results from a step-like movement in the basement. (d) Distribution of distortional strain energy resulting from step-like movement in basement (after Sanford).

experiments, the sand was uncompacted, and in others it was compacted at approximately 15 lb/in².

Typical results obtained by using this apparatus are shown in Fig. 41a, b, c. The deformation and fault development which result from uplift along a broad curve is shown in Fig. 41a. This experiment simulates the conditions for group I, together with the mirror image represented in this analysis. The first fracture to form as a result of this type of displacement occurred as a vertical tension crack at the crest of the fold. This was followed by a series of normal faults at, or near, the crest of the fold. Additional faults sometimes developed at progressively greater distances from the crest as deformation continued. The average dip of these normal faults was approximately 65°.

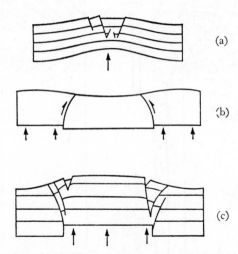

Fig. 41. Results of sandbox experiments due to (a) sinusoidal uplift in basement, (b) Graben step-movement in basement, (c) hoist step-movement in basement (after Sanford).

The kind of results obtained from Horst and Graben type differential movement in the basement are shown in Fig. 41b and c. Each step in the basement represents the analytical conditions postulated in group II.

In the Horst experiments it was found that tensile cracks at the upper surface and the shear fractures at the lower boundary developed simultaneously. As displacement at the basement continued, one or more shear fractures (reverse faults) propagated to the surface while, at the same time, the tension cracks opened and became deeper. This was followed by the development of normal faults in the vicinity of the tension cracks.

It will be noticed that the reverse faults obtained in these experiments are curved fractures. The degree of curvature, it was found, was influenced by the thickness of the layer (the thinner the layer, the greater the degree of curvature of the fault plane), but was independent of the relative movement of the basement blocks (compare Fig. 41b and c).

If it is assumed that for sand, $\varphi = 28°$; then the coincidence between the analytically and experimentally determined fractures is almost exact.

It is interesting to note that the formation of a Graben in the basement, which could well result from a slight extension, gives rise to reverse faults at the surface which could easily be interpreted in the field in terms of a phase of horizontal compression.

CRITICISM OF USING THE BRITTLE FAILURE CRITERIA IN THE FAULTING PROBLEM

The use and application of brittle failure criteria in the interpretation of fault development has been criticized by Odé. In particular, he argues that the Griffith theory demands that planes of failure propagate at a velocity cluse to the speed of sound and that in developing, they completely alter the stress field which initiated fracture. Further, it is suggested that fault patterns, which are produced in sand during experiments of the kind which have been described in previous sections of this chapter, should not be used to substantiate and support analyses based on elastic deformation followed by brittle fracture. It is pointed out that faults in sand and clay in model experiments can be made to propagate as slowly as the experimentor wishes, and can be stopped at any time by reducing the applied forces. In addition, Odé emphasizes that during some triaxial compression experiments specimens shear without losing cohesion. Such evidence leads him to suggest that faulting is a manifestation of velocity discontinuities in a mass undergoing plastico-viscous deformation.

It may be inferred from the time–strain mechanism considered in Chapter 1 that the criticism put forward by Odé regarding the speed of propagation of a plane of brittle failure is not a valid one. Admittedly, brittle failure in short-term experiments in the laboratory often occurs with explosive violence. Nevertheless, with care, the speed at which brittle failure planes propagate, in laboratory experiments, can be controlled. For example, velocities of propagation of cracks in glass ranging from 2 to 200 ft/sec have been measured by Edgerton and Barstow; other examples will be quoted later. Further, the speed of development of a Griffith crack can, on theoretical grounds, be related to the applied stress. Thus, Mott has shown that the velocity of propagation $\dot{c}$ of a crack can be written in terms of the stress σ at the tip of the crack, as follows

$$\dot{c} = \sqrt{\frac{2\pi V}{k}} \left[1 - \left(\frac{\sigma_c}{\sigma} \right)^2 \right]^{\frac{1}{2}}) \tag{81}$$

where k is a constant, V is the velocity of sound in the material and σ_c is the critical stress needed to cause propagation of the crack. It may be inferred from this equation that the speed of propagation will be very slow for values of σ which are only infinitesimally greater than σ_c. Poncelet reaches similar conclusions. Further, Roesler has pointed out that a condition of quasi-equilibrium between surface and mechanical energy can probably be maintained over long periods. Hence, a slowly moving crack is quite consistent with a slow change in surface energy of the crack.

It is recognized that the boundary conditions which obtain prior to faulting will alter as a result of the development of the shear planes. However, once such faults have developed they introduce a degree of anisotropy into the problem. So that, provided the change in the boundary conditions is not too drastic, it is likely that further movement will take place along these existing planes of weakness rather than develop new fault planes with orientations only slightly different from the earlier ones.

Thus, while there is little doubt that a great many, even the majority of, faults are the result of plastico-viscous deformation, it is highly probable that the development of many faults in competent material, at shallow depths in the earth's crust can validly be interpreted in terms of the brittle failure criteria dealt with in Chapter 1.

Note (see p. 92).

The concept put forward by Rubey and Hubbert of a rectangular prism thrust along, or sliding down a gentle slope, over strata in which the pore-water pressure is very high has been modified by Raleigh and Griggs [1963, *Geol. Soc. Am. Bull.* v. 74. Effect of the Toe of a Thrust Plane, p. 819]. In this paper the authors consider the influence of various forms of "toes" on the behaviour of the prism-shaped block for the type of condition postulated by Rubey and Hubbert, thereby making the whole analysis far more realistic and feasible. The effect of the various types of toes is to reduce somewhat the length of block which may be thrust along a horizontal plane or to increase by a very small amount the angle down which a block will slide. The presence of a toe at the front of the thrust block obviates the criticism presented on pages 86 and 91, for now the uniaxial strength of the rock in the thrust block in no way sets a limit upon the thickness of the block.

CHAPTER 3

JOINTS

INTRODUCTION

Joints are cracks and fractures in rock along which there has been extremely little or no movement. They are the most commonly developed of all structures, since they are to be found in all competent rocks exposed at the surface. Yet, despite the fact that joints are so common and have been studied widely, they are perhaps the most difficult of all structures to analyse. The analytical difficulty is attendant upon a number of fundamental characteristics of these structures. Thus, there is abundant field evidence that demonstrates that joints may develop at practically all ages in the history of rocks. In sedimentary rocks, for example, joints may develop soon after deposition, while the sediments are still unconsolidated. They may possibly develop towards the end of a phase of active tectonic compression, and be associated with faults and folds. Or they may develop much later, when the phase of active deformation has subsided. Moreover, tectonic deformation is not necessary to the development of joints, for competent rocks which exhibit no evidence of tectonic deformation are cut by joints.

In the light of these observations, it is unlikely in the extreme that all joints are the result of a single mechanism.

Another difficulty in joint analysis springs from the fact that, characteristically, joints exhibit little or no displacement along the joint plane. Consequently, except in special instances, it is extremely difficult, even impossible, to establish the age relationship of joint planes with one orientation to those with a different orientation. As a result, incorrect assumptions regarding the ages of joints may easily be made and this can invalidate the conclusions of the analysis.

Yet another stumbling block encountered when dealing with joints stems from the difficulty in giving a hard and fast definition (which will be acceptable to all geologists) of what constitutes a joint. However, although a rigid, universal definition may not be possible, there is a general nomenclature which is in common use in describing these structures. An outline of this nomenclature is given in the following sections.

CLASSIFICATION OF JOINTS

Joints may be classified and described with reference to one or more of a number of their characteristics, such as shape, size and their relative importance which combines size and frequency of occurrence.

Shape

If joints are planar and parallel, or sub-parallel, so that they form *sets*, the joints are said to be *systematic*. When joints are irregular, curved, or conchoidal fractures they are said to be *non-systematic*. Such curved joints are little used in the analysis of joint development. Consequently, whenever the word joint is used and not qualified it will subsequently be assumed that the structure is one of a systematic set of joints. It should be noted that the term systematic joints is used in a sense which is entirely different from the term joint system. The latter term is used to indicate intersecting sets of systematic joints.

Size

The magnitude of joint planes forms a continuum which covers a tremendous range, from structures which are sometimes hundreds of feet in extent down to microscopic sizes. The subdivision and classification according to size of such a continuum is arbitrary and will tend to vary according to the user and the magnitude of the largest joints in the area under investigation. As we shall see later, the size of joint planes is largely related to the lithology and the size of the rock units being studied.

Joints which cut through a number of beds or rock units and which can be traced for many tens, or hundreds of feet (provided the exposure is sufficiently extensive) represent one end of the size-scale and are termed *master joints*. Joint planes which are an order of magnitude smaller but which are still well defined structures may be called *major joints*. Smaller, relatively unimportant breaks may be called *minor joints*. Finally, at the lower end of the size-scale there are the minute fractures which sometimes occur in finely banded sediments of varying lithology. These *micro-joints* (or micro-cleats in coal) which may be a small fraction of an inch in extent, grade downward in size into the truly microscopic range.

Frequency and Size (relative importance)

Joint frequency is a term used to indicate the number of planes of one particular joint set encountered in a linear traverse at right angles to the joint planes.

In many localities one set of joints is often dominant, being both larger and/or more frequent than joints of other sets in the same locality. The structures of this dominant set are sometimes referred to as *primary joints*. Often, when this term is used, only one other set of joints is developed and structures of this set are known as *secondary joints*. These terms refer only to the degree of development and do not have a genetic connotation. In particular, the former term should not be confused with the primary joints encountered in igneous rocks which are discussed in a later section.

Joints frequently occur in relatively narrow zones, in which one joint is replaced *en echelon* by another joint which is slightly off-set.

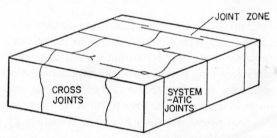

Fig. 42. Block diagram showing systematic joints arranged in zones, and replacing each other *en echelon*, together with non-systematic cross joints.

At their ends, joints are sometimes bifurcated and sometimes linked to the adjacent joint in the zone, as indicated in Fig. 42.

It is commonly observed that the rock between zones is cut by cross joints. (These structures should not be confused with fractures of the same name which develop in igneous rocks.) Hodgson (1961a) maintains that cross joints differ from typical systematic joints, in that cross joints do not intersect systematic joints or well-developed bedding surfaces and although cross joints may sometimes form planar surfaces, they are typically somewhat sinuous and non-systematic.

The descriptive and non-genetic terms noted in the preceding paragraphs are based upon the geometry and relative degree of development of the joints, both individually and in sets. The nomenclature relating to their orientation and relationship to other geological structures is considered in the following section.

RELATIONSHIP OF JOINTS TO OTHER STRUCTURES

Joints and their orientation with respect to other structures have been widely studied in the field and it has been established that systematic joints usually show well-defined relationships to folds and faults which develop during the same tectonic cycle. A general synthesis of the orientation of joint sets to folds and faults is represented in Fig. 43, 44 and 45. It should be borne in mind that in any specific field example the systematic relationships shown in these figures may need to be modified, in that one or more of the sets of joints may fail to develop.

The orientation of joint sets relative to folds is apparently dependent upon the size and type of fold, the relative competence of the rock units in which the structures are formed and the magnitude of the joint planes in relation to the size of the fold and thickness of the rock units.

The relationship between master joints and relatively minor folds is indicated in Fig. 43a. The orientation of some of the joint sets can be related directly to, and defined in terms of the *a-*, *b-*, *c*-axes of the "tectonic cross"; where it is assumed *a* is the direction of movement of the fold, *b* is parallel to the fold axis and *c* is perpendicular to the *ab* plane. The set of joints which cuts the fold at

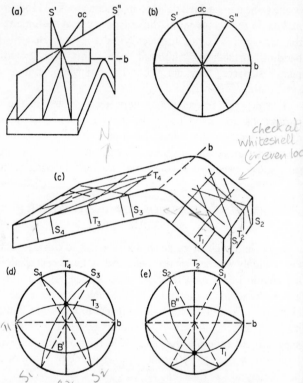

FIG. 43.   (a) Block diagram showing typical relationship of master joints to an anticline. (b) Stereogram of master joints shown in (a). (c) Block diagram showing typical relationship of joints in the limbs of an asymmetrical anticline. (d) Stereogram of joints in the gently dipping limb. (e) Stereogram of joints in the steeply dipping limb.

right angles to the fold axis is classified as *ac-joints*. (The term cross joint has also been used to describe these structures, but its use is not recommended because of the possibility of confusing these *ac-*

joints with the non-systematic cross joints or with the primary cross joints which develop in igneous rocks.) The set which is orthogonal to the *ac* set is known either as *longitudinal* or *bc-joints*. Again,

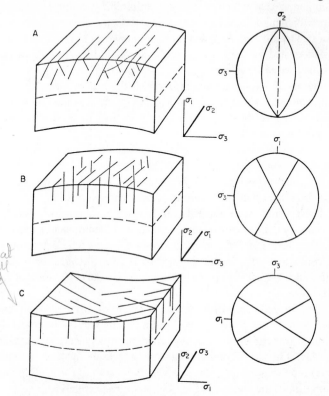

ck at
well
ine

FIG. 44. (a) Joints at the crest of a major anticline with the orientation of normal faults. (b) Shear joints at the crest of a major anticline with acute angle made by the joints intersected by the axial line. (c) Shear joints in the trough of a major syncline (above the neutral surface) with obtuse angle made by the joints intersected by the axial line. The orientation of the principal stresses when these structures developed and stereograms indicating their orientation are also shown.

these structures should not be confused with the primary longitudinal joints which may develop in igneous rocks. The joint sets represented by the planes marked S_1 and S_2 cannot readily be related to the tectonic cross and are sometimes referred to as *oblique joints*.

It will be noted that the oblique joints have an orientation which is similar to that which would be exhibited by conjugate wrench faults which might develop as a result of the same compression which gave rise to the fold. As a result of this similarity in the orientation, it is suggested on dynamic grounds, that the oblique joints can be classified as *shear joints*.

The *ac*- and *bc*-joints which intersect the angles formed by the complementary shear joints are less obviously classified as *tension joints*. The mechanisms which may give rise to such shear and tension joints are discussed in a later section.

It is suggested that the type of joint orientation represented by Fig. 43a is best developed in folded rocks which, at the time of joint formation, varied little in competence from one rock unit to another as, for example, in certain metamorphic suites.

Master joints may frequently fail to form in folds which develop in relatively thin, interbedded competent and incompetent materials. It is then found that the major and minor joints, which develop in the competent units in the limbs, remote from the crest or trough of the fold, are usually strongly influenced by the orientation of the rock unit.

The idealized disposition of the various joint surfaces which may develop in the steeply dipping, "leading" limb and the more gently inclined "trailing" limb of an asymmetrical anticline are indicated in Fig. 43c. The stereographic diagrams (which are in the horizontal plane) representing the joint-sets which develop in the leading and trailing limbs are shown in Fig. 43d and e respectively.

As in the previous example, there are two sets of shear joints and two sets of tension joints. But instead of being related to the fold as a whole, these joints are related to the limbs of the fold.

It will be noted that only the tension joints T_2 and T_4 are vertical. These, of course, have orientations which are identical with each other and with any *ac* master joints which may develop. All other joint planes are inclined to the vertical. They are, however, perpendicular to the surfaces of the rock units in each limb. Since the various joint planes are perpendicular to the bedding, it follows that the line of intersection of the shear joints is also perpendicular to the bedding.

In the field, this ideal relationship may not be exactly realized. Nevertheless, joints of this type are usually not more than 15° from perpendicular to the surface of the rock units, i.e. in sedimentary rocks, perpendicular to the bedding planes. In instances when the joints are not exactly perpendicular to the bedding they commonly tend towards the vertical.

The type and orientation of joints which develop near the crests and troughs of folds depends upon the thickness of the rock unit and the degree of deformation, as indicated be the sharpness of the flexure.

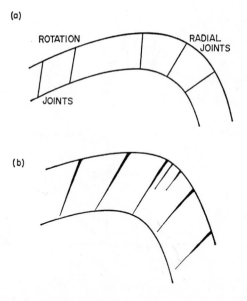

Fig. 45. (a) Section of crest of anticline showing typical orientations of radial joints, about the crest, and rotation joints in the gentle dipping limb. (b) Similar open or quartz or carbonate filled structure sometimes described as "joints".

Some of the joint systems which may develop in thick units of gently folded rock are indicated in Fig. 44. The sets of joints

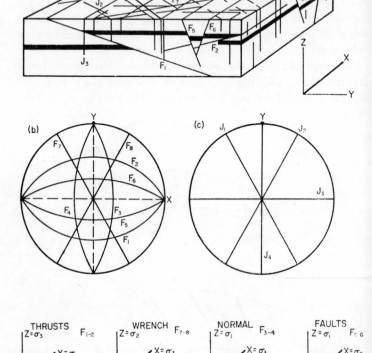

FIG. 46. (a) Block diagram showing orientation of faults and joints in unfolded rocks which may result from various phases of compression and tension related to one complete tectonic cycle. (b) Stereogram of faults orientations shown in (a). (c) Stereogram of joint orientation shown in (a). (d–g) Orientation of stress fields when the various groups of faults were initiated.

represented in Fig. 44a (after Bucher) have the orientation of normal faults with their line of intersection parallel to the fold axis. These joints may be interpreted as shear fractures. It was indicated in Chapter 2, that during the formation of such conjugate shears the axis of intermediate principal stress σ_2 is parallel to the line of intersection of the planes, while the axis of maximum principal stress bisects the acute angle formed by the shears. Consequently, the stress system which must have existed at the time of their formation is as indicated. The vertical stress was the greatest principal stress, the intermediate principal stress acted parallel to the axis of the fold and the least principal stress was in the direction usually associated with the direction of maximum compression.

Shear joints of the type represented in Fig. 44b and c have been described by de Sitter. It will be noted that the orientation of the shear joints in Fig. 44b is different from that of the shear joints represented in Fig. 43a for the axis of the fold intersects the acute angle between the complementary joint sets. Consequently, it may be inferred that at the time of the development of these joints the maximum principal stress acted parallel to the fold axis, and the intermediate stress was in the vertical direction.

The shear joints which may form in the troughs of gently flexured synclines are indicated in Fig. 44c. These planes, it will be seen, are similar in orientation to the master shear joints shown in Fig. 43a.

In folds where the curvature of the crests is sharp, fractures of the type indicated in Fig. 45a may develop. Fractures with a similar orientation are shown in Fig. 45b. However, these fractures are open or, more usually, filled with quartz or some carbonate material, and the fracture surfaces are often rough and somewhat irregular. These gashes, or fissures, almost certainly developed during the process of folding and represent failure following plastic deformation. It is suggested, for reasons which are discussed later, that such structures should not be classified as joints.

The faults and related joints which may develop in a horizontally bedded series of sediments are represented in block diagram a and stereographic diagrams b and c of Fig. 46. This figure represents the types of faults which may develop throughout the various phases of a tectonic cycle. Although the relative positions of the greatest, intermediate and least principal stresses may interchange, it is assumed, for the sake of simplicity, that they are always orientated in the x-, y- and z-directions indicated in Fig. 46a. The orientations of the principal stresses with respect to these directions, during the formation of the various faults are given in Fig. 46d–g.

It will be seen that the wrench faults F_7 and F_8 and the joints J_1 and J_2 are the only structures which have the same general orientation. As noted earlier, in view of the fact that these structures are parallel, or sub-parallel, joints J_1 and J_2 have been classified as shear joints.

It is significant that joints do not, in general, form parallel to the other planes of shear failure, i.e. thrusts and normal faults.

The other sets of joints, J_3 and J_4, which are also vertical fracture planes, bisect the angles formed by the shear joints and are classified as tension joints.

If folds had formed during the compressive phase, it is likely that the axes of these structures would be parallel to the line of intersection of the thrust planes, i.e. the b-axis would be approximately parallel to the x-direction in Fig. 46. Consequently, the J_3 joint set represents the bc-joints and J_4 will represent the ac set of joints.

The normal faults indicated in Fig. 46a are inclined to the vertical at about 30°. However, in the field, it is sometimes found that normal fault planes are near vertical and that they are parallel to one or other of the sets of tension joints. When such a relationship exists, it is often difficult to be sure whether the tension joints formed parallel to the normal faults, or whether the joints predate the normal faults which subsequently developed along these existing planes of weakness.

SURFACE FEATURES

In the absence of well-defined tectonic structures, it is extremely difficult to give a genetic classification of joint systems. However, it has been suggested that the general appearance of the joint surface may be used to differentiate between shear and tension joints. Shear joints are usually markedly planar fractures which are not affected by local changes in lithology. For example, they tend to cut across pebbles in conglomerates, mud pellets, etc., without change of direction of the joint plane. Tension joints in some areas are more irregular surfaces which tend to be deflected by, and follow the outline of, the type of minor variation in lithology noted above.

Another possible means of differentiating between shear and tension joints has arisen from the study of surface features which sometimes develop on the faces of joint planes.

A detailed study of the individual joint planes was first made by Woodworth and more recently by Hodgson (1961a and b) and Roberts. They have observed that joint planes are frequently complicated in their morphology, and a classification of the various parts of the joint plane and the features which develop on the joint surfaces have been evolved.

A schematic block diagram (after Hodgson) showing the primary surface structures of a systematic joint is seen in Fig. 47. When the joint is unweathered, the dominant feature, known as the *main joint face*, usually has a slightly rough or granular surface on which faint ridges or rays may form a pattern. The most commonly observed pattern, indicated in Fig. 47 and Plate 2, is known variously as a *barb*, *plume* or *feather structure*. (It is thought advisable to refer to this pattern as a plume structure rather than a feather structure since this term may be confused with the feather joints associated with faults.) The axis of the plume on the main joint face is commonly parallel to the upper and lower surfaces of the rock unit, although, very occasionally, the axis is roughly perpendicular to the bedding.

Another type of feature which may be seen on the surface of main joint faces consists of circular, or near circular, concentric ridges, as shown in Plate 2 and Frontispiece. These patterns often develop in coal and have been termed "augen" fractures by German miners. (These surface features are not to be confused with the augen structures which sometimes develop in metamorphic rock.)

In some instances, the main joint face is separated from the surrounding *border* or *fringe*, by a pronounced *shoulder* which, where present, is usually formed by an abrupt termination of the fine ridges of the plume structures on the main joint face against *en echelon* small-scale joints which develop in the fringe. These *fringe joints* (*f-joints*) are usually set at an angle of 5–25° to the main joint face. The change in strike from the main joint to the fringe joints may be gradual or abrupt, while the direction of off-set of the *f*-joints relative to the main joint face may be either left- or right-handed. When plume structures develop in the *f*-joints, they tend to be orientated at high angles to the boundaries of the rock unit. The main joint with such surface features is usually a shear fracture. It seems probable, therefore, that the *f*-joints either represent minor tension joints or the complementary shear fracture to the main joint face.

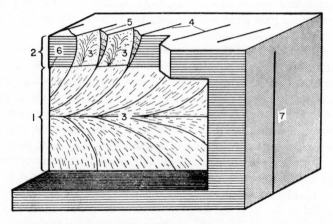

FIG. 47. Details of surface features on a joint plane (after Hodgson). 1, main joint face; 2, fringe; 3, plumose structure; 4, *f*-joints (B-planes); 5, *c*-fractures; 6, shoulder; 7, trace of main joint face.

Occasionally, small *cross-fractures* (*c-fractures*) may be observed between the *f*-joints, which are analogous to the larger cross joints.

Plume structures were recorded by Parker when he studied the joint systems in areas of New York State. He reported that they were well developed upon joint planes which, from other evidence, were classified as shear (compression) joints, but that they were infrequently developed upon the joint planes which were attributed to tensile failure.

Roberts reached a similar conclusion from his study of joints associated with a minor fold. Two sets of joints which were symmetrical to, and subtended an angle of 30° with, the *ac* plane of the fold (i.e. shear joints) exhibited abundant plume structures, while the *bc*-joints (i.e. tension joints) displayed no tendency to develop such structures.

Surface marks, very similar to those shown in Fig. 47 and Plate 2, have been obtained in the laboratory on a wide range of materials including rock, metal, glass, gelatine and other substantially isotropic substances.

Surface marks which have developed in glass have been studied by Murgatroyd. It was found that markings fell into two main groups, described as *rib-marks* (which are equivalent to the ridges around augen structures) and *hackle-marks* (which are similar to the elements making up plume structures).

A cross-section taken through rib-marks (see Fig. 48a) indicates that an individual rib is actually a high point where a fracture, which has been moving upward, comes to rest, perhaps for the merest instant, then resumes its course in a downward direction. The directions, upward and downward, refer to the section shown in

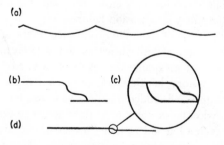

Fig. 48. (a) Cuspidal section of one type of rib-mark. (b) Section through single hackle-mark. (c) Similar section showing join up of adjacent surfaces. (d) Indicating relative relief of the hackle-mark compared with the extent of the surface on either side.

Fig. 48a of course, and not to the movement of the developing fracture in the field.

In a series of experiments, Murgatroyd demonstrated that rib-marks always developed on very slow-moving fractures, while, in his experiments, fractures which developed at relatively high speeds were always smooth faced. The association of rib-marks with slowly developing fractures may also be inferred from the fact that rib-marks frequently occur in "fatigue" fractures (Den Hartog).

Murgatroyd concluded that the experimental results indicate that the forces which cause the arrest and change in direction of the fractures are molecular and not due to the readjustment of the applied force. It is suggested that the arrest in movement of the developing fracture plane may possibly be attributed to shock waves generated by the formation of the crack. Poncelet suggests a similar mechanism.

It may be noted in passing that the direction of propagation of the fracture can be determined from these markings, because the ribs invariably present the convex face to the direction of movement.

It seems probable that there are at least two types of rib-marks. The section of rib-marks shown in Fig. 48 are cuspid. In the Frontispiece, however, it will be seen that the flanks of the rib-marks are planar. The acute angle made by these planes is approximately 30°, and it is tentatively suggested that the one flank represents a shear fracture while the other may either represent the complementary shear, or else it may be a tension failure surface.

It may also be noted from the Frontispiece that hackle-marks are occasionally associated with rib-marks, so it is likely that these latter structures are not invariably the result of slow-moving fractures; for in the laboratory, hackle-marks are commonly associated with explosive rupture and sudden fracture due to relatively large forces.

Murgatroyd obtained sections at right angles to the trend of the hackle-marks (see Fig. 48b) which shows that they are escarpments which make an abrupt transition from one plane to another. A diagrammatic section through adjacent hackle-marks is shown in Fig. 48c. (The true order of relief and relative amplitude of these structures are indicated in Fig. 48d.) It will be noted that the hackle-marks show a striking similarity to a joint zone in which individual systematic joints are linked to the adjacent off-set joints by non-systematic cross-joints.

Carlson, in a study of failure and crack propagation in metals, concludes that plume structures (he calls them *herring-bone marks*) develop as a result of micro-flaws forming in advance of a fast moving main fracture plane which, as it propagates, incorporates each micro-flaw as individual hackle-marks. However, he emphasizes that this mechanism applies only to materials in the semi-brittle state; for, he claims, if the material is truly brittle, the fracture surfaces are smooth and free from plume structures.

Since hackle-marks have been produced on surfaces which have developed perpendicular to the axis of least principal stress, as, for example, in "Brazilian", or "Disc" tests (see the many excellent photographs published by Gramberg) there is, as yet, some doubt whether one may designate field structures as either tensile or shear joints solely on the presence, or absence, of plume structures.

Nevertheless, these surface structures provide an insight into the mode of joint development. For, by analogy, the development of large joint planes is often the result of concomitant spreading of a whole series of linking micro-fractures. As they develop, these fractures release energy in the form of shock waves, which sometimes influence the subsequent development of the fracture plane and may result in the formation of rib-marks. It will be noted that these conclusions are completely in accord with the concept of brittle, or semi-brittle failure, discussed in Chapter 1.

Moreover, if Carlson's findings may be applied to rock, and if the previous conclusion that plume structures are associated with shear joints and not with tension joints is accepted as valid, then it follows that, for any one rock unit, when shear joints formed, the rock was in a semi-brittle state, but was in a brittle state when tension joints developed. Hence, it may be inferred that tension joints formed at a higher level in the crust than shear joints. This is a conclusion which is arrived at by theoretical means in a later section of this chapter.

Finally, it may be noted that hackles, like rib-marks, can also be used to determine the direction in which a fracture travelled, for the barbs always point in the opposite direction from the movement of the developing fracture.

JOINT MOVEMENT

Since joints are fractures, it follows that some movement of the adjacent rock mass has taken place. In the case of tension joints the movement has been perpendicular to the joint planes, while for shear joints it is to be anticipated that some degree of differential movement has occurred along the joint plane. The question which obviously arises is how much movement is permitted before a joint grades into some other form of structure?

In freshly exposed unweathered surfaces (as, for example, in a mine or a drill core) the joint trace is often a "hair-line" crack, which only opens and becomes more apparent after drying out or slight weathering has taken place. In some instances, fractures are lined or filled with quartz or some carbonate material, which may range from a small fraction of an inch to many inches, or even feet, in thickness. In dealing with such structures, it is obviously important to record and report the thickness of the filler material. When it exceeds a small fraction of an inch (say one-tenth of an inch, although this is, of course, a completely arbitrary limit), it is suggested that the structure be referred to as a vein, or dyke, rather than a joint.

An example of two quartz-filled tension joints cut and displaced by a later, dolomite-filled shear joint is shown in Plate 3. The differential movement of the shear joint is approximately one-hundredth of an inch.

The plume structures which occur on many shear joints are fine structures with little relief, and any significant differential movement along the joint would completely obliterate them. Consequently, it may be concluded that even master shear joints, which exhibit such surface features, have suffered an exceedingly small differential displacement. Indeed, if a planar fracture shows evidence of differential movement which can be detected in the field, or has slickensided surfaces, it is probably preferable that the fracture be regarded as a minor fault.

It will be appreciated that the definition of a joint as a planar fracture, along which there has been only an infinitesimal amount of movement, is not unique; for example, it applies equally well to fracture cleavage. Willis realized this and suggested that fracture cleavage represents the condition in which joint planes are exceedingly closely spaced. This is a viewpoint to which many geologists do not subscribe. Turner considers that cleavage, schistosity and other "s-surfaces" should not be confused with, or classified with joints. Such "s-surfaces", it is suggested, are the result of plastico-viscous deformation and should not, on genetic grounds, be grouped with joints which are almost certainly the result of brittle failure.

ORIGIN OF JOINTS

A number of hypotheses have been advocated to account for the development of joints. Some of the earliest ideas, which are no longer given credence, related the formation of joints to magnetic forces or to cleavage in minerals. It has been suggested that joints form as a result of earthquakes, or that they developed in response to tensile stresses which resulted from contraction of sediments, folding or from regional compression.

Because of their orientation with respect to other structures, it has also been suggested that some joints are shear fractures which develop as a result of compression. Torsion has also been invoked to explain the origin of joints.

Each of the hypotheses proposed can be criticized on some grounds, and it is quite clear that no single mechanism can be held responsible for the development of all joints.

Consider, for example, the concept that joints are tensile fractures. The lack of movement parallel to the joint face is evidence in support of this hypothesis. It is difficult, however, to envisage that all joint sets are the direct result of tensile stresses acting perpendicular to the joint planes, for it would be necessary to postulate a complex and changing stress system to account for each set of joints.

One of the criticisms which can be levelled against the theory which suggests that many joints are the result of shear failure during compression was put forward by Kendall and Briggs. They pointed out that while the compression and shear theory explains the orientation and development of faults, it does not offer any satisfactory reason for the large number of joints which develop. Moreover, the straightforward compression and shear hypothesis cannot readily explain the lack of displacement and differential movement along the shear joints.

Instead of compression and shear, Daubree, who based his ideas on experiments with thin sheets of glass, advocated a torsional theory of joint formation. In a series of classic experiments he produced sets of "joints" which intersected at approximately right angles. Bucher, in criticizing Daubree's interpretation, points out that the glass failed in tension at the upper and lower faces of the sheet. Hence, the intersection of the "joint" sets was more apparent than real.

Later, Kendall and Briggs modified the torsional theory of joint formation and suggested that joints were fatigue phenomena, and that they resulted from alternating, regularly orientated torsional stresses brought into play by semi-diurnal tidal action in the solid rock, caused by the gravitational attraction of the moon. Joints, they concluded, developed early in the history of sediments and are the result of minute, rhythmic, differential movements along the joint planes which already existed in more competent sediments underlying the unconsolidated units. These movements, in time, resulted in the upward propagation of joint planes into the unconsolidated sediments. Mollard suggests that this propagation of joints from older to newer sediments through fatigue may be augmented and reinforced by occasional earthquake shocks.

Certainly, there is evidence which shows quite conclusively that joints can, and do, develop in recently deposited sediments. For example, Kinahan describes jointing in "recent", poorly consolidated, or unconsolidated, sediments which was comparable with jointing developed in older consolidated sediments. Joint sets which intersect at right angles in the clays which were formally the bed of the Great Salt Lake have been reported by Gilbert, while Crosby describes joints which have developed in the "wet" unconsolidated

Miocene clays and sands which form cliffs along the western shore of the Potomac River, Virginia. One set is approximately parallel with the cliff face and a second set is almost at right angles to it.

It is, however, difficult to envisage that the joints in such un-consolidated sediments could survive the pressures which obtain during the processes of compaction and consolidation when the sediments are buried, often at considerable depths.

A similar objection has been raised by Turner regarding the development of joints in metamorphic rocks. He argues that no rock undergoing plastico-viscous deformation can, at the same time, develop joints; for rock flow is apt to eradicate any jointing the rock may possess. Consequently, it seems certain that, in metamorphic rocks at least, joints formed after the main tectonic compression, when the rocks were no longer plastic. From this, it may be inferred that, in general, joints in metamorphic rocks developed at higher levels in the crust than those at which the rocks underwent tectonic deformation. The present writer believes that this argument also applies to a high percentage of joints which have developed in consolidated and competent sedimentary rocks.

It has been shown that joints often form an integral part of "movement pictures". Hence, it must be concluded that the orientation of the principal stresses, which obtained during the main tectonic phase, must have been closely related to that which existed during the formation of these "post-compression" joints. There is, however, one important difference. Many joints, especially in horizontally bedded series, are vertical, or near vertical, fractures, and where these are classified as shear joints this implies that during the development of these joints, the intermediate principal stress σ_2 acted in the vertical direction. But it has been shown (see Chapter 2) that the conditions which must be satisfied before the inter-mediate principal stress acts in the vertical direction are rather special and, judging from field evidence, infrequently met. However, shear joints are widespread in their occurrence and are found in areas where there are no wrench faults in evidence. Consequently, in the light of these remarks, the existence of the ubiquitous shear joint presents a problem.

A mechanism was advanced by the author (Price, 1959) which overcomes this difficulty; for it explains the existence of the large numbers of tension and shear joints and their orientation with respect to other structures, and also explains the lack of differential movement exhibited by shear joints. The mechanism was based on the assumption that rocks could retain residual strain energy, which assumption was subsequently proved to be valid (Chapter 1). It was then suggested that the residual stresses associated with the residual strain energy were modified during uplift in such a way as to give rise, either to tension joints on their own, or, in certain cases, to both shear and tension joints. The argument presented was as follows.

During the main tectonic phase, the resistance to deformation of the rocks builds up until it is in equilibrium with the maximum available tectonic stresses. When these conditions are attained, all further changes in strain and deformation cease. If the rocks are ideal Bingham bodies (and it has been shown in Chapter 1 that they approximate to this concept in the upper levels of the crust), then the strain energy and the associated residual stresses will remain stored in the rock. While there is no further strain the residual stresses will faithfully represent in quantity and direction the stresses which acted at the end of the tectonic phase of active compression.

Joints are observed in rocks which are encountered at the surface or in mines at relatively shallow depths. Consequently, it may be assumed that following the tectonic deformation there was, in general, a phase in which the rocks were lifted up, and superincumbent material removed by erosion and denudation. This uplift and decrease in cover is attended by a number of effects. Firstly, there is a decrease in the gravitational loading. Secondly, there is a decrease in the lateral stress which is related to the value of m (Poisson's number). This, in turn, as we have seen, is related to the stress environment, so that the lateral stresses due to this effect would vary as shown in Fig. 29. There is a third factor which causes variation in the stress field, and this is related to the lateral strain which develops in rock during uplift.

Consider this last point. A unit of competent rock which is horizontal and unfolded can be represented in section as an arc of radius R from the centre of the earth. If uplift takes place without tilting of the beds, the radius becomes $(R + dR)$ where dR is the amount of uplift. From Fig. 49 it is quite clear that as a result of the uplift there has been an increase in the lateral extent of the uplifted rock unit. At AC it had a length L, while at BD (after uplift) it has a length $(L + dL)$.

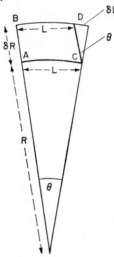

Fig. 49. Indicating extension of bed length L to $L + dL$ after uplift from AC to BD.

Now it is clear from Fig. 49 that

$$L = R \cdot \theta$$

and

$$dL = dR \cdot \theta.$$

Consequently, the lateral strain in the bed which results from uplift is given by

$$dL/L = dR/R. \qquad (82)$$

If it is assumed, for the sake of convenience, that dR, the actual amount of uplift, is 20,000 ft, or approximately 4 miles, then, since R is approximately 4000 miles

$$dL/L = dR/R = 1/1000.$$

This lateral stretching is represented in Fig. 49 as "plane-strain". The strain in the earth's crust due to this effect will, of course, be radial in the horizontal plane.

If this horizontal expansion is brought about purely by elastic stretching, a tensile stress σ_T tends to develop, so that

$$\sigma_T = E \, . \, dL/L \qquad (83)$$

where E is Young's modulus.

It has been shown that the value of Young's moduli for rock varies with the rock type and with the confining pressure. For the two competent coal measure sandstones, the data for which are represented in Fig. 29, the modulus parallel to the bedding is in excess of $1 \times 10^7 \text{lb/in}^2$. From the discussion presented in Chapter 1 it may be inferred that at stresses below the fundamental yield strength of the rock, the error incurred by assuming that the long-term elastic modulus is equal to the short-term modulus is negligible. Also, it will be appreciated that the rocks are still in compression; consequently, one may use these values rather than the value of Young's modulus in extension.

Therefore, using these values in eqn. (83) it follows that the tensile stresses which would tend to develop in these rocks if they experienced uplift of 20,000 ft is approximately 10,000 lb/in².

Now the gravitational load in the vertical direction increases by approximately 1 lb/in² for every foot of increase of depth of cover. Thus, in an uplift of 20,000 ft, the vertical gravitational load decreases by approximately 20,000 lb/in². Therefore, the tensile stresses which tend to develop due to the horizontal extension of the beds during this amount of uplift are equal to half the change in the gravitational load. This relationship is, of course, only approximate. But for the sake of simplicity and convenience, it is assumed to hold for the majority of competent rocks at the depths being considered.

By using this approximate relationship, it is now possible to consider the variations in the stresses which occur in horizontally bedded rocks as a result of uplift. Two examples will be considered: (a) it will be assumed that the initial conditions approximate to the hydrostatic state, and in (b) the initial condition represents the final phase of compression in which the rocks retain residual stresses.

(a) Initial Condition Approximately "Hydrostatic"

Consolidated sedimentary rocks, which exhibit no evidence of tectonic compression when observed in outcrops, are seen to be cut by joint planes. How did these joints develop? To answer this question, it is necessary to consider briefly the stresses which may develop during the various stages of the rock's history.

It has been noted in Chapter 2 that when sediments are covered by subsequent deposits the stresses which develop may be very close to the hydrostatic state. During burial the sediments may be compacted, and thereby attain their ultimate physical properties, such as strength and elasticity. If uplift of such sediments takes place without the development of significant horizontal, tectonic compressive stresses, then one can assume that the initial stress condition is close to the hydrostatic state. The initial stress condition and the stresses which subsequently develop as the result of uplift† are indicated diagrammatically in Fig. 50.

Point B represents the stress values obtaining in all directions at the maximum depth of burial. The straight line AB represents the decrease in the gravitational load during uplift. This decrease is, of course, directly proportional to the amount of uplift that takes place. The distance OC represents the value of the horizontal stress which would obtain if this were a completely elastic problem, i.e. if it were assumed that the rocks always possessed their final elastic properties. However, it has been assumed that the final elastic properties of the rocks under consideration were attained at depth O. The curve AGC has been inserted merely to indicate the manner in which a portion of the lateral stresses will decay during uplift. This component of the lateral stress is related to the change in vertical gravitational load, and the variations of the values of Poisson's number m; the curve AGC is taken from Fig. 29. As uplift takes place, the lateral component of stress due to gravity

---

† In this analysis it is assumed that concomitant erosion occurs and keeps step with the uplift so that the original surface datum remains unchanged.

decreases in a manner similar to that indicated by curve *AGC*, so that the decrease in the total lateral stress due to this factor is represented by *BD*. A further decrease in the lateral stresses is brought about by the lateral stretching due to uplift. As noted earlier, it is assumed that this tensile stress σ_T is equal in value, but opposite in sign, to half the decrease in vertical load which occurs during uplift. Thus, the resultant variations in the lateral stresses which take place as a result of uplift are represented by *BEF*.

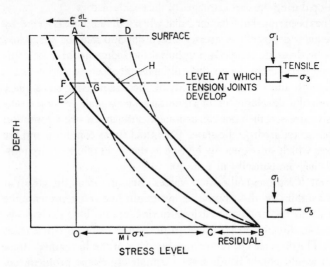

Fig. 50.   Variations in stress due to uplift from an initial condition of near hydrostatic stress.

It will be seen that the total lateral stresses decrease more rapidly than the vertical gravitational stress, so that the vertical stress becomes the greatest principal stress. These are the conditions conducive to the formation of normal faults. It will be noted, however, that the maximum depth of cover represented in Fig. 50 is approximately 20,000 ft, so that the initial hydrostatic pressure approximated to 20,000 lb/in². The maximum stress difference $\sigma_1 - \sigma_3$ is represented by the line *FH*, and is less than 10,000 lb/in².

Such a stress will, in general, be insufficient to cause shear failure in competent rocks of the type under consideration.

At point E, the lateral stresses are zero, and at shallower depths the rocks actually go into tension. The estimated fundamental, long-term tensile strength of competent rocks is of the order of 250 to 1000 lb/in², so that further uplift very readily results in the formation of vertical tension joints. In the unlikely event that the tensile stresses are exactly equal in all horizontal directions, the joints which develop may be randomly orientated.

In general, however, it is to be expected that the initial stress conditions will only approximate to the hydrostatic state. The lateral stresses will not necessarily be equal, so that, on uplift, one will represent the least, and the other the intermediate, principal stress. King-Hubbert and Willis have shown experimentally that, in such a stress field, tension joints, when they develop, form at right angles to the axis of least principal stress. Consequently, when the stress conditions indicated by point F in Fig. 50 are attained, i.e. when the tensile stresses equal the tensile strength of the rock, joints form which are perpendicular to the axis of least principal stress.

The moment the fractures develop in the rock, the tensile stresses are relieved; the erstwhile least principal stress changes, and is probably replaced by a compressive stress which can have a maximum value of $\sigma_x/(m-1)$, cf. point G. The other horizontal principal stress, formally the intermediate principal stress, becomes the least principal stress. Further uplift can eventually cause a second set of tension joints to develop; the two sets forming an orthogonal system.

If the lateral stresses were initially very similar, the formation of the two sets of joints may be practically simultaneous. Indeed, the formation of the first set may even trigger off the development of the set which, it is tentatively suggested, may develop as cross-joints of the type shown in Fig. 42.

(b) Initial Condition with Residual Tectonic Stresses

If it is assumed that the tectonic stresses which develop during a compressive phase remain as residual stresses, then, assuming also that lateral compression acted in one direction without concomitant relief perpendicular to the compression, the stress conditions prior to uplift are those represented by eqns. (46) and (47) in Chapter 2. Consequently, the lateral stress conditions which arise during uplift are given by

$$\sigma'_y = \frac{1}{m-1} \cdot \sigma_z + c_y - \sigma_T \qquad (84)$$

and

$$\sigma'_x = \frac{1}{m-1} \, \sigma_z + \frac{1}{m} \cdot c_y - \sigma_T \qquad (85)$$

where, as before, $\sigma_T = E.\, \mathrm{d}L/L$ [eqn. (83)] are the tensile stresses which result from the lateral stretching of the beds during uplift.

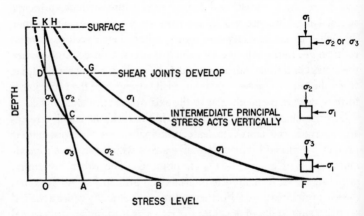

Fig. 51. Variations in stress due to uplift from an initial condition in which the ratio of greatest to least principal stress σ_1/σ_3 is large.

The initial and subsequent stresses which develop during uplift are represented in Fig. 51. The initial condition represents the stresses which are residual after the development of thrusting. The greatest principal stress which, at this stage, acts horizontally is represented by OF. As uplift takes place, the greatest and intermediate principal stresses change in the manner indicated by curves FG and BCD respectively. The reasons for these changes have been discussed in the previous example. At the level indicated by C, the vertical load changes from the least principal stress to the intermediate principal stress. Thus, one of the conditions which must be fulfilled before vertical shear fracture can develop, is satisfied. At this stage, the ratio of greatest to least principal stress may be too small to cause such fractures to develop. However, it can be inferred from Fig. 51 that, as uplift progresses further, the ratio of the greatest to least principal stress increases rapidly until the second set of conditions necessary for the development of shear fractures is fulfilled at, or near, the level indicated by *DG*.

As soon as shear joints develop, an unknown proportion of the residual stresses is released. The stress system which then obtains is indeterminate. The development of the shear fractures will tend to reduce the maximum principal stress, and increase the least principal stress. The vertical load due to gravity may, at this stage, assume the role of the greatest principal stress. These conditions are, of course, similar to those discussed in the previous example. Further uplift may cause the rock to pass into tension with the possible subsequent development of a system of tension joints.

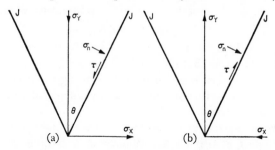

Fig. 52. (a) Normal and shear stresses on shear joints *JJ* when the compressive stress bisects the acute angle and the tensile stress bisects the obtuse angle made by the joints. (b) Normal and shear stresses on shear joints *JJ* when the tensile stress bisects the obtuse angle and the compressive stress bisects the acute angle made by the joints.

However, it is distinctly possible that the development of subsequent lateral strain could, under certain circumstances, be accommodated by opening of the recently formed shear joints, or by differential movement along the joint plane.

Since the relative values of the principal stresses are not known, it is necessary to revert to the designation of the axes of principal stresses in terms of x, y and z, where, as before, σ_z is the vertical principal stress and σ_x and σ_y are the horizontal principal stresses.

If the vertical stress is the greatest principal stress, this is the condition which would ordinarily give rise to normal faults. However, as noted earlier, the stress differential is insufficient to cause shearing to occur. Consequently, attention can be restricted, as a first approximation, to the horizontal stresses.

Initially, it will be assumed that, following the development of shear joints, the principal stress σ_y is greater than the other horizontal principal stress σ_x, and that, as a result of further uplift, there is a tendency for σ_x to become tensile.

The conditions envisaged are represented in Fig. 52a. It will be seen that the normal stress σ_n acting across a shear joint will be given by

$$\sigma_n = \sigma_y \sin^2 \theta - \sigma_x \cos^2 \theta \qquad (86)$$

where σ_y is compressive and σ_x is tensile.

Clearly, σ_n will be greater than zero and compressive when

$$\sigma_y \geqslant \sigma_x \cot^2 \theta = \sigma_x K_1 \qquad (87)$$

where $K_1 = \cot^2\theta$. If the ratio σ_y/σ_x is less than K_1, the normal stress will be zero or tensile and there will be a tendency for the tensile stress σ_x to be dissipated by opening the existing shear joints, or, since there will be no frictional resistance, by differential movement along the joint plane.

However, it is possible that shear movement may take place along the shear joints, even if the normal stress σ_n is compressive. If it is assumed that the joints have no cohesion, movement will take place along the pre-existing joint plane when

$$\tau = \mu_s \cdot \sigma_n. \qquad (88)$$

The normal stress acting across the joint plane is given in eqn. (86) and the shear stress which develops along the plane is given by

$$\tau = (\sigma_y + \sigma_x) \sin \theta \cdot \cos \theta. \qquad (89)$$

Hence, movement will tend to occur when

$$(\sigma_y + \sigma_x) \sin \theta \cos \theta = \mu_s (\sigma_y \sin^2 \theta - \sigma_x \cos^2 \theta),$$

which may be expressed in terms of σ_y, as

$$\sigma_y = - \sigma_x \frac{\cos\theta\,(\sin\theta + \mu_s\cos\theta)}{\sin\theta\,(\cos\theta - \mu_s\sin\theta)}. \qquad (90)$$

However, if it is assumed that the angle of internal friction of the rock φ_i is equal to the angle of sliding friction φ_s, then θ and μ_s are interrelated, for $2\theta = \pi/2 - \varphi$, where $\varphi = \tan^{-1}\mu$. If this relationship is substituted for μ_s in eqn. (90), it can be shown that the expression resolves to

$$\sigma_y = - \sigma_x \cot^2\theta. \qquad (91)$$

Thus, from eqns. (87) and (91), it follows that there is always a tendency for the tensile stress to be dissipated by opening or by differential shear movement along the shear joint planes.

Consequently, the formation of tension joints perpendicular to the σ_x axis of principal stress will tend to be inhibited.

It is emphasized that this simple analysis does not take into account the effect of the frictional forces which will act along the bedding planes and tend to oppose the dissipation of the horizontal tensile stresses which would be brought about by the differential movement along the shear joints. The restraining influence of the frictional forces acting on the bedding plane will be related to a large number of variables, such as lithology, thickness of bed, inclination of the bed and depth of burial, and cannot readily be estimated. However, it seems reasonable to assume that, eventually, the tensile stress σ_x will in part be dissipated by differential slip along the shear joints, and in part will give rise to a set of tension joints.

In the example considered in the previous paragraphs, σ_x has the same orientation as any "b" lineations which may have developed during the main phase of active compression. Hence, any tension joints which develop normal to the σ_x axis of stress may be classified as ac-joints.

Either movement along the shear joint planes, or the development of tension joints will result in the horizontal stresses being altered. The stress σ_x may now become compressive and have a maximum value of $\sigma_z/(m - 1)$. To further the analysis, let us assume that, with further uplift, the stress σ_x becomes and remains compressive but that the other principal stress σ_y becomes tensile, as shown in Fig. 52b. It may be inferred that if σ_n is to remain greater than, or equal to zero, then

$$\sigma_x \geqslant \sigma_y \tan^2 \theta = \sigma_y K_3. \tag{92}$$

As before, if the ratio of the stresses σ_x/σ_y is greater than a factor K_2, the normal stress σ_n will remain compressive.

The possibility also exists, as in the previous example, that the tensile stress may be relieved by movement along the shear joint planes while the normal stress is compressive. Again, if it is assumed that the shear joints exhibit no cohesion, then shear movement will take place when the conditions represented by eqn. (88) are attained; so that movement will take place when

$$(\sigma_y + \sigma_x) \sin \theta . \cos \theta = \mu_s(\sigma_x \cos^2 \theta - \sigma_y \sin^2 \theta)$$

which may be expressed in terms of σ_x/σ_y as

$$\frac{\sigma_x}{\sigma_y} = -\frac{\sin \theta \, (\mu_s \sin \theta + \cos \theta)}{\cos \theta \, (\sin \theta - \mu_s \cos \theta)}. \tag{93}$$

It will be noted that if these conditions are to be satisfied while the normal stress remains compressive and σ_y is tensile, then shear movement can only take place when $\tan \theta$ is equal to, or greater than μ_s. If, as in the previous instance, it is further assumed that the coefficient of friction μ_s, and θ are interrelated, then the condition $\tan \theta > \mu_s$ is only satisfied for values of θ between 30° and 45°.

The negative sign σ_y has been taken into account in eqn. (93), so that for values of θ from 30° to 45°, σ_x/σ_y is always greater than $-K_2$. Consequently, the stresses will tend to be relieved by differential movement along the existing joint planes.

For angles of θ less than 30°, shear movement along the shear joints (when the normal stress acting across the joint planes is greater than zero and σ_y is tensile) is not possible. Furthermore, σ_x need only be a small fraction (0 to 0·33) of the value of the tensile stress σ_y to maintain a compressive normal stress across the shear joint planes. Hence, it is unlikely that the tensile stress will be relieved by opening, or by differential movement along the shear joints.

It may be inferred from this simple and approximate analysis, therefore, that the degree of development of tension joints is related to the angle of intersection of the shear joints. When $\theta = 45°$ (i.e. the angle of intersection of the shear joints is 90°) it is to be expected that the degree of both ac and bc tension joints will be similar, and will be related to the frictional restraint which exists along the bedding planes. However, it is usual for shear joints to intersect so that 2θ is considerably less than 90°, and it is often 60° or less. For such shear joints (when $2\theta < 60°$), the analysis indicates that the bc tension joints are more likely to develop than the ac-joints. The smaller the value for the acute angle of intersection of the shear joints, the more pronounced will be the differences in the degree of development between the two sets of tension joints. As far as the author is aware, there exists no systematic study of the relationship between the degree, or frequency, of development of ac- and bc-joints relative to the angle 2θ. However, the joint development in folded sediments in South Pembroke (Hancock) lends support to the idea that tensile stresses may cause movement along earlier formed shear joints rather than bring about the development of fresh tension joints. Thus, one system of shear joints formed in these structures have the usual orientation with respect to the fold axis, i.e. the acute angle between the complementary joint sets, which in this instance is only 40°, is symmetrical about the a-direction. In this area the ac tension joints have failed to develop, but the left-handed joints of the above mentioned shear system frequently contain in-filling material. Hence it may be concluded that during a phase when extension was parallel to the fold axis, the stresses were dissipated by opening of pre-existing shear joints rather than by the development of a fresh set of ac tension joints.

JOINT FREQUENCY

In addition to the point brought out in the previous section, there are other aspects of joint frequency which are worthy of note. For example:

1. Joint frequency is commonly many hundreds of thousands of times greater than "fault frequency".

2. Joint frequency, in any one locality, is not a constant, but varies with the lithology of the rock type.

3. For any single lithological type, joint frequency is related to the dimensions of the rock unit.

4. Joint frequency is influenced by the degree of tectonic deformation.

These points are considered in the following paragraphs.

1. *Joint and Fault Frequency*

It is suggested that the difference between joint and fault frequency is related to the degree of movement of the two structures and to the different stress environments which obtained when they developed.

Movement along fault planes has the effect of shortening a faulted block in the direction of the greatest principal stress and increases the extent of the block in the direction of the least principal stress. The shearing movement along the plane may be considerable; consequently, the stress conditions initiating this movement may be relieved over a wide area.

It has been proposed in an earlier section that many, even the majority of, joints in competent rock are post-compressional structures, and that the state of stress which obtains prior to the formation of shear joints is different from that which gives rise to faulting. The stresses which bring about the development of joints, it has been suggested, are residual and are not replenished by tectonic processes. The shearing stresses at the instant of joint development are probably of the order of a few tens of thousands of pounds per square inch. Such stresses can, in general, be dissipated by a movement of only a small fraction of an inch along a shear plane. Such a movement would, however, only dissipate the residual stresses in the immediate vicinity of the joint plane. In order to dissipate the stresses throughout a wide area, the formation of a very large number of joints is necessary.

2. *Joint Frequency and Lithology*

The most striking relationship between joint frequency and lithology is probably that which exists between coal and adjacent sediments. For example, a 1 ft. thick coal seam usually has well-defined joint, or *cleat*, planes with a separation of a fraction of an inch, while in an adjacent sandstone of similar thickness, the joint planes will probably be a foot or more apart. Price (1959) has suggested that the number of joints which develop in a rock is related to the strain energy originally stored in the rock. Now the strain energy stored in an elastic body is equal to the work done in producing a given amount of strain. Thus, in a unit cube which has a linear stress–strain relationship, the strain energy w is given by

$$w = \tfrac{1}{2}\sigma \cdot \varepsilon \qquad (94)$$

where σ is the applied stress and ε is the resulting strain. But $\varepsilon = \sigma/E$ (where E is Young's modulus). Therefore

$$w = \sigma^2/2E. \qquad (95)$$

From eqn. (44) which relates to the conditions of strain in triaxial stress, and eqn. (95), it can be shown that the strain energy stored in a unit cube subjected to triaxial compression is given by

$$w = \frac{1}{2E}\left[\sigma_1{}^2 + \sigma_2{}^2 + \sigma_3{}^2 - \frac{2}{m}\left(\sigma_1\sigma_2 + \sigma_1\sigma_3 + \sigma_2\sigma_3\right)\right]. \qquad (96)$$

From this equation it will be seen that even if the stresses in the adjacent rocks are the same, the strain energy which these rocks contain is related to the elastic "constants" E and m, and may, therefore, be very different. It is interesting to note that if one assumes typical values of E and m for coal ($E = 2 \times 10^5$ lb/in^2 and a mean value of $m = 3$) and a strong coal measure sandstone ($E = 1 \times 10^7$ lb/in^2 and a mean value of $m = 5$) and further assume that these rocks are both subjected to a vertical load of 10,000 lb/in^2 and a maximum horizontal compression of 20,000 lb/in^2; then it can be shown that the strain energy stored in the coal is 32·5 times greater than that stored in the sandstone. Thus, the ratio of joint frequency and the strain energy under these postulated conditions is of the same order.

However, it must be pointed out that the postulated stress conditions are not the ones likely to give rise to joint development, since the decrease in strain energy which would take place during uplift has been neglected. Further, it is not known at what stage of compaction and compression the rocks attained their final elastic properties. Consequently, the good agreement between joint frequency and the estimated strain energy may be to some extent fortuitous.

3. *Joint Frequency and Bed Thickness*

From detailed measurements carried out on sediments in two areas of Wyoming, Harris *et al.* concluded that for a given lithological type, the concentration of joints is inversely related to the thickness of the bed. For example, in one specific locality, joints in a 10 ft. thick dolomite bed occur at every 10 ft.; while in a 1 ft. thick dolomite bed from the same locality, the joint frequency was one every foot. Similar relationships between mean joint separation and bed thickness have been obtained for two: sandstone and a limestone by Kirollova, Novikova and Bogdanov (see Fig. 53).

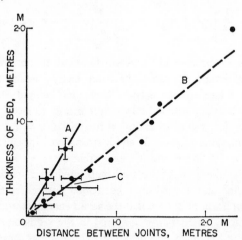

Fig. 53. Relationship between bed thickness and distance between joints (from data by Bogdanov, Kirollova and Novikova).

It is suggested that this relationship between thickness of bed, or rock unit, and joint frequency can be related to the frictional forces which exist between adjacent beds. Consider a single horizontal competent rock unit set between two relatively incompetent beds (see Fig. 54a). Assume that the competent bed has a distributed tensile stress σ_T acting throughout the layer, and is everywhere on the point of failing in tension; and that this tensile stress causes the

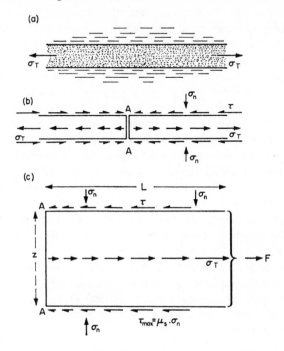

FIG. 54. (a) Uniform tensile stress σ_T acting in a single competent unit. (b) Indicates the reduction in tensile stress due to the formation of a single joint coupled with the development of shear stresses along the bedding planes which prevent excessive opening of the joint. (c) Details of stress intensity in section of competent bed length L and thickness Z.

development of the first tension joint at AA (Fig. 54b). Clearly, if the competent bed were completely free to move, this joint would continue to open until the tensile stresses were relieved over a wide area. However, it is obvious from Fig. 54b that, even if it is assumed that there is no cohesion between beds, a frictional shearing stress will develop along the bedding planes so as to oppose widening of the tension joint.

Since the joint forms a free surface, the horizontal stress in the competent unit at the joint plane itself, is reduced to zero, but gradually increases in intensity away from the joint plane. The value of this tensile stress in the vicinity of the joint at any given distance, is related to the sum of the traction forces which act along the surface of the competent bed. The original tensile stress σ_T is reduced near the joint plane by a minute amount of differential slip or differential strain along the bedding planes until the sum of the horizontal forces in the competent bed, at any point, exactly equals the sum of the traction forces along the surfaces of the rock units. At some distance L from the joint AA, the traction forces are sufficient to maintain the original stress σ_T. At this distance, which represents the limit of influence of joint AA, the tensile stress is sufficiently large to form a second tension joint. Hence, the distance L represents the minimum distance at which a second joint can develop and will obviously be closely related to the mean joint separation.

The important feature of this concept is that the frictional traction resisting bedding plane slip must balance the total force acting along the competent bed. In a simple example (see Fig. 54c) the total force acting along the bed (here assumed to be of unit width) is given by

$$F = \sigma_T \cdot \zeta$$

where ζ is the thickness of the bed. It is assumed that the frictional traction acting along the bedding planes over a distance L is just sufficient to balance the total force F. Clearly, if the bed is twice as thick as that represented in Fig. 54c the total force, acting along the competent unit, when the stress is σ_T is doubled. However, the normal stress and the coefficient of friction between the competent and incompetent units remain unaltered. Consequently, the traction must act over a distance $2L$ before it is sufficient to balance the total tensile forces acting within the competent bed.

This simple and approximate mechanism, in which it has been assumed that the cohesion between adjacent beds is non-existent and that μ, σ_n and σ_T are all constant, gives rise to an exact relationship between joint separation and bed thickness. In actual fact, of course, there may be varying degrees of cohesion between adjacent strata. Similarly, the coefficient of friction and tensile strength will vary according to relatively minor changes in lithology, while the normal stress will be influenced by depth of cover, orientation of the rock unit and, if the bed is inclined to the horizontal, the degree of lateral compression. Nevertheless, it may be inferred from the data given in Fig. 53 that in any one locality these factors cause only second-order variations in the relationship between joint frequency and bed thickness.

4. *Joint Frequency and the Degree of Tectonic Deformation*

It has been noted that the degree of susceptibility of any stratum to joint development is largely controlled by the thickness and lithological characteristics of the stratum. Consequently, in order to investigate the influence of tectonic deformation in joint frequency, it is necessary to deduce the factors relating joint frequency to lithology and bed thickness, and then reduce the data to some standard reference datum. To do this, it is recommended that a datum bed is chosen which has a widespread distribution throughout the area under investigation, and which also possesses a uniform lithology. The joint frequency in adjacent beds can be related directly to this datum bed. In turn, the joint frequency in the adjacent beds can be related to the joint frequency in beds more remote from the datum bed.

To determine the bed thickness factor, the lithology is kept constant. For example, one compares joint frequencies in dolomite beds. To determine the lithology factor, on the other hand, one compares the joint frequency in beds of different lithology, but which all have the same thickness. Once these factors have been evaluated, they may be used to convert the joint frequency data in beds of different lithology and thickness to the datum bed. From this corrected data, a map can be produced showing the distribution of the intensity of joint development and its relationship to other structures.

These methods were applied by Harris *et al.* to two areas in Wyoming, the relatively small Goose Egg Dome and the more extensive flexure in the Sheep Mountain area. It was found that the highest joint frequencies were associated with areas of maximum curvature of the structures, i.e. in areas where the rate of change of either dip or strike of the sedimentary units was greatest.

It will be noted that this conclusion is completely in keeping with the concept expressed earlier, that joint frequency is fundamentally controlled by the strain energy content of the rock.

JOINTING IN FOLDED ROCK

For the most part, it has hitherto been assumed that the principal stresses during the main compression and subsequent uplift acted in the vertical and horizontal directions. This assumption was made purely for the sake of simplicity in presenting the argument relating to the development of post-compression joints.

However, in Chapter 2 it was shown that the stress trajectories during deformation are sometimes inclined to the horizontal and vertical directions. This orientation is particularly likely to obtain during the development of belts of asymmetrical folds.

It is reasonable to assume that the orientation of these inclined stress trajectories will be reflected by the orientation of the post-compressional master joints which may subsequently develop. Hence, master joint planes which cut through relatively minor asymmetrical folds may be inclined to the vertical.

A detailed discussion of the mechanics of formation of major and minor joints associated with folds should be based upon a knowledge of the orientation and relative values of the stresses which develop in folded rocks. Unfortunately, little is known regarding the stress systems which may be induced during folding.

One of the many difficulties encountered in attempting an analysis of the stresses which develop in folded sediments is that the physical properties of the rock will, in general, change as the folding takes place. In the majority of instances, especially if deformation is slow and high pore-water pressures do not develop, the sediments will become progressively compacted and strengthened as the fold evolves, only attaining their final physical properties when folding has ceased.

In certain instances, however, sediments will have become indurate, by the deposition of interstitial siliceous or carbonate cement, prior to folding. An indication of the stress systems which can develop in such rock types near the crest or trough of a fold may be obtained if it is assumed that the regional compressive stress, which gives rise to the folding, is locally modified as a result of simple elastic bending.

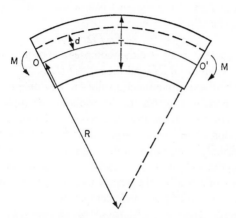

FIG. 55.   Portion of beam, thickness T, bent to a radius R by bending moment M.

Consider a simple beam of the type shown in Fig. 55, which is bent into an arc by a bending moment M; the elastic strain which develops in the beam during bending is given by

$$\varepsilon_b = d/R \qquad (97)$$

where d is the distance of any element from the line OO', and R represents the radius of curvature of the line OO'. Along the line

OO', which represents a curved plane known as the neutral surface, the strain due to bending is zero. Above the neutral surface the strain, which is tensile for an anticlinal flexure, increases in proportion to the distance of the element of the beam from the line OO' and reaches a maximum at the upper surface of the beam. Similarly, below the neutral surface, the compressive strains increase to a maximum along the lower surface of the beam.

The stresses which are associated with the strain at the upper surface of the beam are given by

$$-\sigma_u = E \cdot T/2R \tag{98}$$

where T is the thickness of the beam and E is Young's modulus. Hence, for flexures of small amplitude, the stress σ'_u at the upper surface of an indurate, competent bed near the crest of an incipient anticline is the sum of the regional compressive stress σ_r and the stresses due to bending; so that

$$\sigma'_u = (\sigma_r - E \cdot T/2R). \tag{99}$$

Clearly, the stress σ'_u may be either compressive or tensile, depending upon the values of σ_r, E and the ratio $T/2R$.

It is not possible to give precise values to any of these symbols; nevertheless, certain estimates may be made. For example, the value of E for the cemented siliceous or calcareous sandstones, or indurate limestones being considered, will probably fall within the range of 10^6 to 10^7 lb/in^2. The value of the regional stresses which are likely to give rise to folding will depend upon a number of factors, such as depth of burial of the sediments, the regional pore-water pressure and the thickness and competence of the dominant members in the succession. At moderate depths (say 10,000 to 20,000 ft) the value of the regional compressive stress acting along the competent units will probably fall within the range of 10^4 to 10^5 lb/in^2.

If these values are substituted in eqn. (99), it will be seen that for very modest ratios of $T/2R$ (say 1/50 to 1/100), when gentle flexures are formed, the stresses in the upper elements of the competent beds near the crests of anticlines will be tensile. As the ratio $T/2R$ becomes smaller (i.e. as folding progresses) these stresses may cause the material to fail, by the formation of one or more tension cracks which will be orientated approximately perpendicular to the upper surface of the rock unit. As folding progresses, the tension gash, or gashes, will spread downward into the bed. If only one gash develops, it is subsequently likely to form the nick point of a chevron fold.

If, instead of fracturing, the rock undergoes considerable plastico-viscous deformation before finally rupturing, the crest is likely to be well rounded. In the event that tensile deformation occurs before the stresses below the neutral surface reach the yield point in compression, the neutral surface will migrate downward as folding progresses. Consequently, the zone of plastico-viscous extension may be quite extensive in the vicinity of a sharply flexured anticlinal crest, and may even spread almost to the lower surface of the rock unit. Throughout this zone, the tensile stresses are likely to be reasonably uniform in magnitude (assuming that rock exhibits little or no strain hardening) for the stresses will be equal to the tensile yield strength of the rock. These stresses will be orientated parallel, or sub-parallel, to the upper and lower surfaces of the rock unit. Consequently, if the rock eventually ruptures, a series of tension gashes will develop which will be similar to the structures represented in Fig. 45b.

If folding ceases before such fractures develop, the tensile stresses around the crest will be unrelieved and will remain as residual stresses. As we have seen, during subsequent uplift, rocks undergo lateral extension, and further tensile stresses are induced which reinforce and augment those already in existence near the crest of the anticline. This build-up in the tensile stresses will, it is suggested, result in the development of radial joints of the type shown in Fig. 45a.

The series of events depicted in the preceding paragraphs were based on the assumption that the elastic moduli and other basic physical properties of the rock were not materially altered by the deformation. In general, however, it seems likely that the majority of sediments become compacted and develop progressively higher values of elastic moduli, strength, etc., as a result of deformation. Nevertheless, the sequence of events outlined above is still likely to hold, except that values of the tensile stresses will tend to be reduced and may not give rise to tension gashes. Indeed, tensile stresses as such may fail to develop during the deformation. Even so, the local reduction in the regional compression near the anticlinal crest may be sufficient to change the relative positions of the greatest, intermediate and least principal stresses. The orientation and relative magnitude of the principal stresses may be inferred from the structures which develop near the crest. For example, the fractures which have the orientation of normal faults in Fig. 44a can be interpreted in terms of the indicated stress system in which the maximum principal stress was approximately vertical, the intermediate principal stress acted parallel to the axis of the fold, and the least principal stress acted in the general directions usually associated with the regional compression. Similarly, the shear structures represented in Fig. 44b developed as a result of compression acting along the axis of the fold and the intermediate principal stress was, in this instance, vertical and the least principal stress, as in the previous example, acted approximately parallel to the regional compression.

One cannot be sure whether joints with the orientation indicated in these two diagrams formed during the folding, or whether they are post-compression structures. However, it is the author's opinion that if fractures with the orientations represented in Fig. 44a and b developed during the active folding of the sediments, the fractures would become normal, or wrench faults. Whereas, it is suggested, joints with such orientations would probably be post-compressional structures which developed during subsequent uplift.

It will be noted that in the trough of the syncline, the stresses above the neutral surface which develop due to bending, are compressional and these reinforce the regional compressive stresses. Clearly, providing cross-folding is absent, the conditions which need to be satisfied before vertical shear planes may develop in horizontally bedded rock must also be satisfied in the trough of a syncline, above the neutral surface. Consequently, the uncertainty of the previous paragraph does not hold and it may be concluded that the shear joints represented in Fig. 44c are very probably post-compression fractures which developed during uplift.

This latter conclusion will also apply to the development of shear joints in uniformily dipping limbs at some distance from the crests and troughs of folds.

In a gently folded series of interbedded competent and incompetent sediments, the disposition of minor thrusts, etc., often indicates that the maximum principal stress is orientated parallel to, and is transmitted along the competent units. It is probable that during subsequent uplift the lateral extension will also tend to be resolved along the competent units. That is, during the period of uplift, the axis of principal stress will tend to be approximately parallel and normal to the surfaces of the rock units. Such a disposition of stress axes would result in the post-compressional joint planes forming at approximately right angles to the bedding.

In the light of this possibility, it follows that when joints in folded rocks are everywhere approximately perpendicular to the bedding, it would be unwise to advance this relationship as sole and positive evidence which purports to show that joint development pre-dates folding. In addition to such evidence, which is not diagnostic, it would, for example, be necessary to demonstrate that joint planes are bent by the folding and that the frequency of development of jointing is completely unaffected by the folding.

When interbedded rocks differ little in competence the stress pattern is likely to be less influenced by the orientation of the rock units. In such instances, the tension joints which develop parallel to the fold axis may have an orientation somewhere between the vertical and the perpendicular to the surface of the rock units. Such structures have been termed *rotation joints* and have been attributed to a shear mechanism resulting from folding.

PINNATE OR FEATHER JOINTS

When joints or fractures abut or intersect, the acute angle (sometimes termed the dihedral angle) is occasionally very small, i.e. 15° or less. In those instances when one of the structures is a fault, smaller subsidiary fractures, which are often termed pinnate or feather structures, will have an orientation similar to the splay fault represented in Fig. 30. Indeed, such pinnate structures may be second-order shears along which differential movement is small. If the stresses are insufficiently large to form second-order shears, pinnate joints with similar orientation to such second-order shears may develop under the influence of residual stresses, at higher levels in the crust.

Such joints may be used to indicate the direction of movement along the main fracture plane, for the acute angle the pinnate joint makes with the primary shear indicates the direction of shear movement along the primary structure.

Joint sets also intersect to form small acute angles. Such angles may occur when one set of shear joints is developed (to the exclusion of the complementary set) and intersects with a set of ac tension joints. However, it may be inferred from Mohr's envelope that conjugate shear joint systems may also form with a small dihedral angle, provided they develop when the least principal stress σ_3 was tensile but not sufficiently great to give rise to tensile failure (Muehlburger).

INTERRELATIONSHIP OF JOINT SETS

In the preceding sections on the mechanics of development of post-compression joints, it has been indicated that in horizontally bedded rocks which have undergone little, or only slight tectonic compression, two sets of tension joints may develop; while in rocks which have undergone considerable tectonic compression, but have remained unfolded, two sets of shear joints may develop, and subsequently these may be followed by two sets of tension joints.

It is emphasized that these structures are related to one single cycle of subsidence—compression and uplift. When a larger number of joint sets are present in horizontally bedded rock units, it is reasonable to conclude, therefore, that the rocks have probably undergone more than one cycle of subsidence and uplift.

Prior to or accompanying uplift, it is possible that the simple sequence of events so far considered may be complicated by regional expansion or compression which may be in a direction which is completely different from the earlier compression. Such extension or compression will then upset the existing stress conditions. For example, extension in the general direction of the least principal stress will facilitate the formation of shear joints; but extension in the direction of the greatest principal stress will impede, or even prevent, the development of such joints, so that only tension joints form. Furthermore, a complete reorientation of the residual stresses is possible so that joints which subsequently develop will be completely unrelated to the movement picture of the original phase of compression.

It is suggested that the general sequence, indicated in the preceding paragraphs, in which shear joints are followed by tension joints, is one of the reasons which makes it so difficult to date the relative joint sets in the field.

When there is more than one cycle of events, tension joints may be cut by shear joints, and a careful examination enables the relative ages of such joints to be determined. Such a relationship is represented by Plate 3.   From field evidence, the earlier quartz-filled tension joints have been attributed to the Caledonian Orogeny, while the later shear joint is probably a post-compression structure which developed during the Hercynian Orogeny (Price, 1962).

The formation of joints has been related in the previous sections to the residual compressive tectonic stresses and the tensile stresses which tend to develop during uplift. This does not, however, negate the possibility that other mechanisms are involved. For example, thermal stresses will be generated, as a result of the cooling of the sediments as they are uplifted, and will reinforce the stresses which result from lateral stretching. Similarly, the rhythmic torsional tidal stresses (Kendall and Briggs) could be very important when the rock is already near the point of failure due to the mechanism outlined above. Moreover, the actual triggering off of the development of many of the joint planes could be the result of shock waves, which may be related to earthquakes, or may equally be related to the development of a nearby joint plane.

It was noted by Dale that, occasionally, when joints developed in granite during the process of quarrying, they gave rise to a dull report as they propagated. Thus, it may be inferred that when a joint forms and spreads it releases energy which makes itself manifest in the form of a shock wave. One may envisage, for example, a shock wave generated by a joint developing in one bed and leaving a trail of hackle-marks (see Plate 2), triggering-off failure in adjacent beds which had already begun to fail, possibly in fatigue, as evidenced by the rib-marks.

Joints which develop in one bed may also propagate upward or downward into adjacent rock units as a result of frictional drag along the bedding planes between the units. It has been suggested that this is one of the mechanisms which may give rise to jointing in incompetent material.

In a sedimentary series containing rocks with different physical properties and varying amounts of strain energy, it is clear that rocks in different rock units may develop at different times. It is distinctly possible that the elastic properties, etc., of the various rock types are so different that the development of joints in the various rock units may be separated by a very considerable period. In the time which elapses following the formation of joints in one rock unit and before the joints form in adjacent rock units, it is possible that the stress system in the unjointed rocks may undergo a slight reorientation. As a result of this, the joints which subsequently develop will have an orientation which is different from those which developed earlier in adjacent competent units. In addition, of course, variations in the orientation of shear joints from one rock unit to another, may also be attributed to differences in the coefficient of friction, in the various rock types, which will affect the angle of shear.

JOINTING IN IGNEOUS ROCKS

The development of primary structures in igneous rock, which have been widely studied by Hans Cloos† and his co-workers, can sometimes be related to the mode of emplacement of an intrusive mass.

---

† An excellent summary of the ideas and conclusions of Cloos *et al.* has been given by Balk.

Cooling and hence crystallization of an igneous melt first takes place at the walls and roof of the intrusive mass. Continued movement and intrusion of the still liquid core gives rise to the development of primary fractures in the solid, but often still plastic, outer shell of the intrusion.

Four main types of primary fractures are recognized and defined with respect to *flow-lines*, *flow-planes* and *Schlieren* which develop during the movement of the viscous liquid melt during the process of intrusion. These are *cross "joints"* (or *Q "joints"*), *longitudinal "joints"* (or *S "joints"*), *diagonal "joints"* and *flat-lying "joints"*. The orientation of these fractures with respect to flow-lines, etc., is represented in Fig. 56.

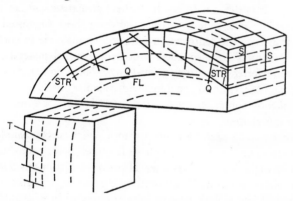

Fig. 56. Orientation of primary joints relative to surface of intrusion and internal structures (after Cloos).

Cross "joints" are among the earliest of fractures to develop in the cooling mass. Typically, they form perpendicular to the flow-lines. In fact, in areas of an intrusion where flow-lines are absent, cross "joints" cannot be identified with certainty; for such "joints" can only be tentatively correlated with cross "joints" which develop elsewhere in the intrusion where flow-lines are present. These fractures are frequently occupied by aplite, or else are almost invariably coated with hydrothermal minerals. The fracture surfaces commonly exhibit slickensiding.

Cross "joints" are regarded as tension fractures which formed when the outer portion of the intrusion had consolidated, as a result of differential movement and drag of the liquid core against the walls and roof; and, possibly, as the result of continued expansion of the intrusion.

Diagonal "joints" form at 45°, or more, to the trend of the flow-lines. Displacement along the fractures indicates that they are shear phenomena which resulted from compression normal to, and extension in the direction of, flow-lines. These fractures are also commonly filled with aplite or hydrothermal minerals.

Primary flat-lying "joints", according to Balk, tend to develop where the apex, or dome, of an intrusion is flat, or in flat sheets and laccoliths. It is difficult to see how these structures can be interpreted on dynamic grounds. It has, however, been suggested that they form when the centre of an intrusion shrinks due to cooling. These structures are also filled with hydrothermal minerals, and have been referred to as primary flat-lying "joints" so that they may be distinguished from barren joints with a similar orientation, which are frequently found in igneous masses and which are dealt with later in this chapter.

Longitudinal "joints" are steep planes which strike parallel to flow-lines. The orientation of these fractures is little affected by variations in pitch of the flow-lines. However, variations in trend of the flow-lines are faithfully followed by changes in the strike of the longitudinal joints. This type of joint is rarely filled with aplite or "dyke" material, and the minerals are usually different from those found in the other forms of primary fractures. Moreover, differential movement of the joint surfaces is rarely observed.

It is suggested that these characteristics indicate that longitudinal joints tend to form later than the other primary fractures. It seems probable that these joints developed in response to tensile stresses which were generated by cooling of the igneous mass, coupled with uplift and lateral stretching.

It will be noted that, with the possible exception of longitudinal joints, primary fractures frequently exhibit evidence of considerable shear movement. Moreover, it appears that the rock mass was in the plastic state when these structures developed. Consequently, it would probably be better to classify these structures as faults and dykes rather than joints. It is for this reason that the present author has used inverted commas when describing these primary structures, to differentiate them from true joints.

However, the stress systems which gave rise to the primary structures may have influenced the development of joints which formed during a later phase, when the intrusion was cooler, brittle, and possibly undergoing slight lateral extension as a result of uplift. Such joints are likely to be barren, but may have an orientation which is closely related to the primary structures.

In addition to primary structures and cooling joints, igneous rocks are almost certain to contain joints which result from regional tectonic compression. Such joints are best identified when they can be related to major faults and shear zones which cut through the igneous mass (Blyth and Firman).

As with the primary structures, it is inferred that in many of the areas described, shearing has taken place while the igneous mass and the country rock have been in the plastic state. The shears and related structures have consequently been related, and likened, to those formed during the classic experiments on wet clay conducted by Reidel, rather than to the dynamic processes of brittle failure.

COLUMNAR JOINTS

The formation of primary structures may, in part, be attributed to cooling of the igneous rock mass. However, *columnar "joints"*, which are such common features of sills and some dykes, are wholly related to the shrinkage of the rock mass during cooling. Typically, the columns are hexagonal in section (although individual columns may be bounded by four, five, seven or even eight joint planes) and have their long axes orientated perpendicular to the upper and lower surfaces of the rock unit.

It will be noted that a regular hexagonal prism is the geometrical form with the greatest number of surfaces which may be placed in juxtaposition with similar hexagonal prisms, so that there are no gaping voids between any of the adjacent columns. The other prisms which are capable of such close packing are rectangular and triangular in section. Thus, it is conceivable that the tensile stresses generated by cooling (which will be equally developed in all directions parallel to the rock unit) could be released by two orthogonal sets of joints, giving rise to rectangular columns, or by three intersecting sets of joints, giving rise to columns with a triangular section.

Other things being equal, it is clear that the quantity of strain energy released from a single column by the development of the joints will be related to its cross-sectional area. A simple calculation will show that the total areas of the joint faces bounding columns with square and equilateral triangular sections are respectively 10 and 20 per cent greater than the area of joint surface enclosing a hexagonal column of comparable cross-sectional area. Hence, it appears that the approximately regular pattern of joints delimiting the columns observed in the field is a manifestation of the principal of least work. That is, the maximum amount of strain energy is dissipated at the cost of the least amount of work utilized in the formation and propagation of fracture planes.

It may be noted in passing that a similar mechanism will control the development of shrinkage cracks in mud, which are also commonly hexagonal. In this instance, shrinkage is, of course, due to drying out of the sediments and not a result of cooling.

SHEET JOINTS

It has been noted that in addition to the primary flat-lying joints, structures may subsequently develop which have a similar orientation and result in a well developed *sheeting* of the intrusion. When such *sheet joints* are closely spaced they are sometimes termed *mural joints*. A feature of sheeting is that in areas of pronounced topography, the sheet joints tend to develop parallel, or sub-parallel to the surface.

Chapman and Rioux carried out a survey of the joint systems in Arcadia National Park on Mount Desert Island off the coast of

Maine. They found that, in general, sheeting was well developed throughout the area in which hornblende granite was exposed. They observed that the frequency of sheet jointing was related to the depth of cover. On the steep slopes of the U-shaped, glaciated valleys, only one or two thick sheeting layers could be seen. On the higher slopes and mountain tops, however, thin sheeting layers were abundant. A typical relationship between the frequency of sheeting, and also the orientation of the sheeting planes with respect to topography, is indicated in Fig. 57. It is suggested that the relationship represented in this figure indicates that sheeting is largely related to pre-glacial topography. In a few localities, however, sheeting has formed parallel to ice-cut surfaces and in these instances are glacial, or post-glacial, in age.

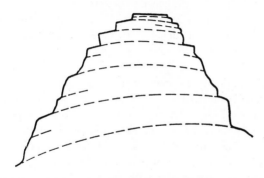

Fig. 57.   Sheet joints, their orientation and intensity in relation to topography (after Chapman and Rioux).

Hill suggests that where the sheeting reflects topography, it is probably caused by a combination of factors, such as expansion of the feldspars and ferromagnesian minerals on weathering, removal of the load of superincumbent rock by erosion, and seasonal variations in temperature affecting the rock near the surface.

It has already been noted that such joints sometimes develop during quarrying and that they form suddenly, emitting a low pitched report. Dale presents other evidence that some granites may contain considerable residual strain energy. In areas of marked topographical relief the maximum and intermediate principal stresses related to the residual strain energy, near the surface, will be orientated approximately parallel to the surface, while the least principal stress will act approximately normal to the surface. Hence, it may be surmised that sheeting joints may be analogous to tension joints described elsewhere, which developed parallel to the axis of maximum principal stress and normal to the axis of least principal stress. In this instance, the least principal stress cannot be expected to be tensile but, at shallow depths, it will be very close to zero. It is possible that the development of the sheets may be likened to an outward buckling of inclined struts. However, it is not understood why sheet joints should form by this mechanism in preference to shear joints which strike parallel to the topography. The factors mentioned by Hill may explain this preference.

Subsequent to the development of sheeting joints, Chapman suggests that other joints may develop. He based this conclusion on the observation that, in many localities, joints terminated against sheeting surfaces. Also, the degree of development and the orientation of these late-forming joints are related to topography. For example, in the Arcadia National Park, as many as six sets of joints were observed in roadside cuttings, whereas a short distance away the number of joint sets was reduced in number. In addition, joints become more abundant and open as cliff faces are approached and joint sets, which have a different trend back from the face, gradually swing around into parallelism with the face, near the cliff edge. Furthermore, observations show that at individual stations the best developed joints appear to fall into a pattern consisting, ideally, of four joint sets; two mutually perpendicular and two diagonal. The diagonal joints are not necessarily at right angles, but are nearly symmetrical with respect to the other two sets. At individual stations, the best developed joints appear to fall into the pattern shown in Fig. 58. The two perpendicular sets conform with the direction of slope and contour of the topography at that particular station. These joints are termed *slope* and *contour joints* respectively.

It is suggested by Chapman that the local joint pattern is best accounted for if it is assumed that slope and contour joint sets formed as a result of downhill sliding of the sheets in response to the forces of gravity. The reason for the development of the diagonal joints is not readily understood, but it is tentatively suggested that they may represent the opening of strongly impressed incipient joint sets which are readily formed. Or, it is suggested, they may be the result of uneven downhill sliding of the sheets due to some local obstructions.

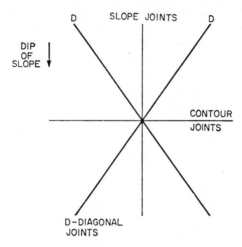

FIG. 58. Typical orientation of Dip, Diagonal and Contour joints
(after Chapman and Rioux).

It may be noted that contour joints are not restricted to igneous rocks, for such structures have been reported in sedimentary rocks by Harris *et al.*

JOINTS—BRITTLE FRACTURES

In geological literature the word "joint" is frequently treated as an omnibus term and has been used to describe structures which vary widely in character and which, in many instances, could better be described as extension gashes, fissures, veins, dykes, minor faults or even cleavage. As a result, the reader is often left in considerable doubt as to the precise type of structure being described when it is merely classified as an "unqualified" joint.

As we have seen, this difficulty is not completely obviated even if one adheres to the definition that systematic joints are fracture planes, normal and parallel to which, movement has been negligibly small. Consequently, it is suggested that genetic criterion be included in the definition of a joint; namely that the term joint be restricted to structures which are the result of brittle fracture.

It is appreciated that the introduction of such a genetic criterion into the classification of geological structures is always fraught with danger. However, genetic criteria have already been introduced by classifying joints on dynamic grounds as shear or tensile fractures. Moreover, the possible errors incurred by the incorrect use of such a genetic criterion can be minimized, for if there is any doubt in the field as to whether a fracture is in fact the result of brittle failure, this doubt can be noted and recorded.

REFERENCES

ADAMS, F. D., 1910, An experimental investigation into the action of differential pressure on certain minerals and rocks, employing the process suggested by Professor Kick; *J. Geol.* **18**, 489.

ADAMS, F. D. and BANCROFT, J. A., 1917, Internal friction during deformation and relative plasticity of rocks, *J. Geol.* **25**, 597.

ANDERSON, E. M., 1951, *The Dynamics of Faulting*, Oliver & Boyd, Edinburgh.

ANDERSON, R. V., 1927, Tertiary stratigraphy and orogeny of the northern Punjab, *Geol. Soc. Am. Bull.* **38**, 665.

ANVRUDDHA, D. and MAXWELL, J. C., 1959, Deformation of quartz grains and sandstones (abs.), *Am. Geophys. Un. Progr. of* 1959, Ann. Mtg., p. 76.

ARTHY, L. G., 1930, Density, porosity and compaction of sediments, *Bull. Am. Ass. Petrol. Geol.* **14**, 1.

BALDRY, R. A., 1938, Slip-planes and breccia zones in the Tertiary rocks of Peru, *Quart. J. Geol. Soc.* **94**, 347.

BALK, R., 1937, Structural behaviour of igneous rocks, *Geol. Soc. Am. Memoir.* **5**.

BALMER, G., 1952, A general analytical solution for Mohr's envelope, *Proc. A.S.T.M.* **52**, 1260.

BLATT, H. and CHRISTIE, J. M., 1963, Undulatory extinction in quartz of igneous and metamorphic rocks and its significance in Provenance studies of sedimentary rocks, *J. Sedimen Petrol.* **33**, 559.

BLYTH, F. G. H., 1950, The sheared porphyrite dykes of South Galloway, *Quart. J. Geol. Soc.* **105**, (3), 393.

BLYTH, F. G. H., 1962, The structure of the north-east tract of the Dartmoor granite, *Quart. J. Geol. Soc.* **118**, (4), 435.

BOGDANOV, A. A., 1947, The intensity of cleavage as related to the thickness of the bed (Russian Text), *Sov. Geol.* **16**.

BOTT, M. P. H., 1959, The mechanics of oblique-slip faulting, *Geol. Mag.* **96**, 109.

BRACE, W. F., 1960, An extension of the Griffith theory of fracture to rocks, *J. Geophys. Res.* **65**, 3477.

BRACE, W. F. and BOMBOLAKIS, 1963, A note on brittle crack growth in compression, *J. Geophys. Res.* **68**, 12, 3709.

BRIDGMAN, P. W., 1949, *The Physics of High Pressure*, Bell, London.

BROWN, C. B., 1938, On a theory of gravitational sliding applied to the tertiary of Ancon, Equador, *Quart. J. Geol. Soc.* **94**, 359.

BUCHER, W. H., 1920, The mechanical interpretation of joints, *J. Geol.* **28**, 707, and **29**, 1.

CADY, W. M., 1945, Stratigraphy and structure of west-central Vermont, *Geol. Soc. Am. Bull.* **56**, 515.

CARLSON, A. J., 1963, On the mechanism of brittle fracture propagation, *Trans. Roy. Inst. Tech. Stockholm Mech. Eng.* **10**, 205, 5.

CHAPMAN, C. A., 1958, Control of jointing by topography, *J. Geol.* **66**, 552.

CHAPMAN, C. A. and RIOUX, R. L., 1958, Statistical study of topography, sheeting and jointing in granite, Acadia National Park, Maine, *Am. J. Sci.* **256**, 111.

CHRISTIE, J. M. and RALEIGH, C. B., 1959, The origin of deformation lamallae in quartz, *Am. J. Sci.* **257**, 385.

CLOOS, E., 1932, Feather joints as indicators of the movements of faults, thrusts, joints and magmatic contacts, *Proc. Nat. Acad. Sci. U.S.A.* **18**, 387.

CLOOS, H., 1936, Plutone und ihre Stellung in Rahmen der Krustenbewegungen, 16th *Int. Geol. Congr. Rep.* **1**, 235.

COTTRELL, A. H., 1959, Theoretical aspects of fracture, *Fracture* (Ed. Auerbach, etc.), p. 20, Wiley, New York.

CROSBY, W. O., 1882, On the classification and origin of joint structures, *Proc. Boston Soc. Natural Hist.* **22**, 72.

DALE, T. N., The granites of Vermont, *U.S. Geol. Survey Bull.* **313**, 354, 404, 484.

DAUBREE, A., 1879, *Geologie Experimentale*, Librare des Corps des points et Chausser des Mines et des Telegraphs, Paris.

DEN HARTOG, 1961, *Strength of Materials*, Dover, London.

DE SITTER, L. U., 1956, *Structural Geology*, McGraw-Hill, New York.

DONARTH, F. A., 1961, Experimental study of shear failure in anisotropic rocks, *Geol. Soc. Am. Bull.* **72**, 985.

DREYER, W., 1955, Über das Festigkeitsverhalten sehr verschiedenartiger Gesteine, *Die Bergbauwissenschaften*, **2**, 183.

EDGERTON, H. E. and BARSTOW, F. E., 1941, Further studies of glass fracture with high speed photography, *Am. Ceram. Soc. J.* **24**, 131.

EIRACH, F. R., 1958, *Rheology*, Academic Press, New York.

FINNIE, I. and HELLER, W. R., 1959, *Creep of Engineering Materials*, McGraw-Hill, New York.

FIRMAN, R. J., 1960, The relationship between joints and fault patterns in the Eskdale granite (Cumberland) and the adjacent Borrowdale volcanic series, *Quart. J. Geol. Soc.*, **116**, (3), 317.

FREUDENTHAL, A. M., 1950, *The Inelastic Behaviour of Engineering Materials and Structure*, Wiley, New York.

GILBERT, G. K., 1882, Post-glacial joints, *Am. J. Sci.* **23**, 25.

GILLULLY, J., 1957, Transcurrent fault and overturned thrust, Shoshone Range, Nevada, *Geol. Soc. Am. Bull.* **68**, 1735.

GOLD, L. W., 1960, The cracking activity in ice during creep, *Can. J. Phys.* **38**, 1137.

GRAMBERG, J., 1961, Axial cleavage fractures and fracture plane analysis, *Mijnbouwkundig Jaarboek*, p. 130.

GRIFFITH, A. A., 1925, The theory of rupture, *1st Int. Congr. Appl. Mech. Proc. Delft.* p. 55.

GRIGGS, D. T., 1936, Deformation of rocks under high confining pressures, *J. Geol.* **44**, 6541.

GRIGGS, D. T., 1939, Creep of rocks, *J. Geol.* **47**, 225.

GRIGGS, D. T., 1940, Experimental flow of rock under conditions favouring recrystallization, *Geol. Soc. Am. Bull.* **51**, 1001.

GRIGGS, D. T., TURNER, F. J. and HEARD, H. C., 1958, Deformation of rocks at 500° to 800°, *Mem. Geol. Soc. Am.* **79**, Rock deformation, p. 39.

HAFNER, W., 1951, Stress distributions and faulting, *Geol. Soc. Am. Bull.* **62**, 373.

HANCOCK, P. L., 1964, The relations between fold and late-formed joints in South Pembrokeshire, *Geol. Mag.* **101**, 174–184.

HANDIN, J. and HAGER, R. V., 1957, Experimental deformation of sedimentary rocks under confining pressure: tests at room temperature on dry samples, *Bull. Am. Ass. Petrol. Geol.* **41**, 1.

HANDIN, J. and HAGER, R. V., 1957, Experimental deformation of sedimentary rocks under confining pressure: tests at high temperature, *Bull. Am. Petrol. Geol.* **42**, 2892.

HANDIN, J., HAGER, R. V., FRIEDMAN, M. and FEATHER, J., 1963, Experimental deformation of sedimentary rocks under confining pressure: pore pressure tests, *Bull. Am. Ass. Petrol. Geol.* **47**, 717.

HARRIS, J. F., TAYLOR, G. L. and WALPER, J. L., 1960, Relation of deformational fractures in sedimentary rocks to regional and local structures, *Bull. Am. Ass. Petrol. Geol.* **44**, 12, 1853.

HEARD, H. C., 1958, Transition from brittle to ductile flow in Solenhofen limestone as a function of temperature, confining pressure and interstitial fluid pressure, *Mem. Geol. Soc. Am.* **79**, Rock deformation, p. 193.

HILL, E. S., 1953, *Elements of Structural Geology*, Methuen, London.

HOBBS, D. W., 1960, The strength and stress–strain characteristics of Oakdale coal under triaxial compression, *Geol. Mag.* **97**, 422.

HOBBS, D. W., 1964 (a) The strength and stress–strain characteristics of coal in triaxial compression, *J. Geol.* **72**, 214.

HOBBS, D. W., 1964 (b) The tensile strength of rock, *Int. J. Rock Mech. Min. Sci.* **1**, 385.

HOBBS, W. H., 1914, Mechanics of formation of arcuate mountains, *J. Geol.* **22**, pts. 1, 2 and 3.

HODGSON, R. A., 1961 (a) Regional study of jointing in Comb Ridge Navajo mountain area, Arizona and Utah, *Bull. Am. Ass. Petrol. Geol.* **45**, 1.

HODGSON, R. A., 1961 (b) Classification of structures on joint surfaces, *Am. J. Sci.* **259**, 493.

HUBBERT, M. K., 1951, Mechanical basis for certain familiar geological structures, *Geol. Soc. Am. Bull.* **62**, 355.

HUBBERT, M. K. and RUBEY, W. W., 1959, Role of fluid pressure in mechanics of overthrust faulting, *Geol. Soc. Am. Bull.* **70**, (1), 115.

HUBBERT, M. K. and WILLIS, G., 1957, *Mechanics of Hydraulic Fracturing*, publ. 415, Shell Oil Co., Tech. Services Division.

INGLIS, C. E., 1913, Stresses in plate due to the presence of cracks and sharp corners, *Trans. Inst. Naval Architects*, **55**, 219.

JAEGER, J. C., 1959, The frictional properties of joints in rock, *Geofis. pur. appl.* **43**, 148.

JAEGER, J. C., 1960, Shear failure of anisotropic rocks, *Geol. Mag.* **97**, 65.

JAEGER, J. C., 1962, *Elasticity, Fracture and Flow*, 2nd Ed., Methuen, London.

JONES, O. T., 1951, The distribution of coal volatiles in the South Wales coalfield and its probable significance, *Quart. J. Geol. Soc.* **107**, 51.

JONES, R., 1958, The failure of concrete test specimens in compression and flexure, *Mech. Prop. of Non-metallic Brittle Materials*, Butterworths, London, p. 243.

KENDALL, P. F. and BRIGGS, H., 1933, The formation of rock joints and the cleat in coal, *Proc. Roy. Soc. Edin.* **53**, 193.

KINAHAN, G. H., 1875, *Valleys and their Relationships to fissures, fractures and faults,* Trubuer, London.

KIROLLOVA, I. V., 1949, Some problems of the mechanics of folding (Russian text), *Trans. Geofian.* **6**.

LENSEN, G. J., 1958, A method of Graben and Horst formation, *J. Geol.* **66**, 579.

LENSEN, G. J., 1959, Secondary faulting and transcurrent splay-faulting at transcurrent fault intersections, *N.Z. J. Geol. Geophys.* **2**, 729.

LIETH, C. K., 1913, *Structural Geology,* Henry Holt.

LONGWELL, C. R., 1922, The Muddy Mountain overthrust in south-eastern Nevada, *J. Geol.* **30**, 63.

LOWRY, W. D., 1956, Factors in loss of porosity by quartzose sandstones of Virginia, *Am. Assoc. Petrol. Geol. Bull.* **40**, 489.

MAXWELL, J. C., 1958, Experiments on compaction and cementation of sand, *Mem. Geol. Soc. Am.* **79**, Rock deformation, p. 105.

MCCLINTOCK, F. A., 1962, On the plasticity of the growth of fatigue cracks, p. 65, *Fracture of Solids* (Ed. Drucker and Gilman), Interscience, New York.

MCCLINTOCK, F. A. and WALSH, J., 1962, Friction on Griffith cracks in rocks under pressure, *Proc. Fourth U.S. Nat. Cong. Appl. Mech.* (Berkley).

MCCONNEL, R. G., 1887, Report on the geological structures of a portion of the Rocky Mountains, *Geol. Surv. Canada Ann. Report,* 1886.

MCHENRY, D., 1948, The effect of uplift pressures on the shearing strength of concrete, *3rd Cong. des Grands Barrages,* Stockholm, C.R.

MCKINSTRY, H. E., 1953, Shears of second-order, *Am. J. Sci.* **251**, 401.

MISRA, A., 1962, Ph.D. thesis, Sheffield University.

MOHR, O., 1882, Über die darstellung des Spannungzustandes eines körpelementes, *Zivil Ingenieure,* **28**, 113.

MOLLARD, J. D., 1957, A study of aerial mossaics in southern Saskatchewan and Manitoba, *Oil in Canada.*

MOODY, J. D. and HILL, M. J., 1956, Wrench-fault tectonics, *Geol. Soc. Am. Bull.* **67**, 1207.

MOTT, N. F., 1948, *Engineering,* **165**, 16.

MURGATROYD, J. B., 1942, The significance of surface marks on fractured glass, *J. Soc. Glass Tech.* **26**, 155.

MURRELL, S. A. F., 1958, Strength of coal under triaxial compression, *Mech. Prop. Non-metallic Brittle Materials,* Butterworth, London, p. 123.

MURRELL, S. A. F., 1962, A criterion for brittle frature of rocks and concrete under triaxial stress, and the effect of pore pressure on the criterion, *Rock Mechanics*, Pergamon Press, Oxford, p. 563.

NOVIKOVA, A. C., 1947, The intensity of cleavage as related to the thickness of the bed (Russian text), *Sov. Geol.* **16**.

OBERHOLZER, J., 1933, Geologie der Glarneralpen Beiträge, *Geol. Karte Schweiz Neue Folg. Lieferung*, **28**.

OBERT, L. and DUVAL, W. I., 1942, Microseismic method of predicting rock failure in underground mining, *U.S. Bur. Min. Rep. Inv.* 3797.

OBERT, L., WINDES, S. L. and DUVAL, W. I., 1946, Standardized tests for determining the physical properties of mine rock, *U.S. Bur. Min. Rep. Inv.* 3891.

ODÉ, H., 1960, Faulting as a velocity discontinuity in plastic deformation, *Mem. Geol. Soc. Am.* **79**, 293.

OROWAN, E., 1949, Fracture and strength of solids, p. 185, *Report on Progress in Physics*, **1**, The Physical Soc., London.

PARKER, J. M., 1942, Regional systematic jointing in slightly deformed sedimentary rocks, *Geol. Soc. Am. Bull.* **53**, 381.

PEACH, B. N. and HORNE, J., 1884, Report on the geology of north-western Sutherland, *Nature*, **31**, 31.

PETTIJOHN, F. J., 1949, *Sedimentary Rocks*, Harper, New York.

PHILLIPS, D. W., 1948, Tectonics of mining, *Mining Magazine*, Sheffield University.

POMEROY, C. and BROWN, J., 1958, Friction between coal and metal surfaces, *Proc. Conf. on Mech. Prop. Non-metallic Brittle Materials*, Butterworth, London.

PONCELET, E. F., 1946, Fracture and comminution of brittle solids, *Am. Inst. Min. Metall. Eng.* **169**, 37.

PRICE, N. J., 1958, A study of rock properties in conditions of triaxial stress, *Proc. Conf. on Mech. Prop. Non-metallic Brittle Materials*, Butterworth, London, 106.

PRICE, N. J., 1959, Mechanics of jointing in rocks, *Geol. Mag.* **96**, 149.

PRICE, N. J., 1960, The compressive strength of coal measure rocks, *Colliery Engng.* p. 283.

PRICE, N. J., 1962, The tectonics of the Aberystwyth grits, *Geol. Mag.* **99**, 542.

PRICE, N. J., 1963, The influence of geological factors on the strength of coal measure rocks, *Geol. Mag.* **100**, 428.

PRICE, N. J., 1964, A study of time–strain behaviour of coal measure rocks, *Int. J. Rock Mech. Min. Sci.* **1**.

REIDEL, W., 1929, Zur Mechanik geologischer Brucherscheinungen. *Zbl. Miner. Geol. Paläent B*. 354.

REINER, M., 1960, *Deformation, Strain and Flow*, H. K. Lewis, London.

ROBERTS, J. C., 1961, Feather-fracture and the mechanics of rock-jointing, *Am. J. Sci.* **259**, 481.

ROBERTSON, E. C., 1955, Experimental study of the strength of rocks, *Geol. Soc. Am. Bull.* **66**, 1275.

ROESLER, F. C., 1956, *Proc. Phys. Soc. (London)* **69**, 55.

RUBEY, W. W., 1951, Structural patterns in the overthrust arc of western Wyoming and adjacent states, *Geol. Soc. Am. Bull.* **62**, 1475.

RUBEY, W. W. and HUBBERT, M. K., 1959, Role of fluid pressure in mechanics of overthrust faulting, Pt. II, *Geol. Soc. Am. Bull.* **70**, 167.

SALMON, E. H., 1952, *Materials and Structures*, **1**, Longmans, London.

SANFORD, A. R., 1959, Analytical and experimental study of simple geologic structures, *Geol. Soc. Am. Bull.* **76**, 19.

SAX, H. G. J., 1946, De tectoniek van het Carboon in het Zuid-Limburgeshe mijngebied, *Meded. Geol. Sticht.*, Ser. C.1.1, Maastricht.

SELDENRATH, R. and GRAMBERG, J., 1958, Stress-strain relations and breakage in rocks, *Mechanical Properties of Non-metallic Brittle Materials*, Butterworth, London.

SERDENGECTI, S. and BOOZER, G. D., 1961, The effects of strain rate and temperature on the behaviour of rocks subjected to triaxial compression, *Proc. Fourth Symp. Rock Mechanics*, p. 83.

SERDENGECTI, S., BOOZER, G. and HILLER, K. H., 1962, Effects of pore fluids on the deformation behaviour of rock subjected to triaxial compression, *Proc. Fifth Symp. Rock Mechanics*, p. 579.

SMOLUCHOWSKI, M. S., 1909, Some remarks on the mechanics of overthrusts, *Geol Mag.* **5**, 204.

SUGGATE, R. P., 1956, Depth-volatile relations in coalfields, *Geol. Mag.* **93**, 201.

TERRY, N. and MORGAN, W., 1958, Studies of rheological behaviour of coal (Conf. Paper), *Mechanical Properties of Non-metallic Brittle Materials*, Butterworth, London.

TERZAGHI, K., 1943, *Theoretical Soil Mechanics*, Wiley.

TIMOSHENKO, S., 1940, *Strength of Materials*, Van Nostrand, N.Y.

TORNEBOHM, A., 1896, Grunddragen af det central Skandinaviens bergbyggnad, *K. Svenska Vetensk Akad. Hanal.* **28**, 201.

TROTTER, F. M., 1949, The devolatization of coal seams in South Wales, *Quart. J. Geol. Soc.* **104**, 387.

TURNER, F. J., 1948, Mineralogical and structural evolution of metamorphic rocks, *Geol. Soc. Am. Mem.* **30**.

VAN WATERSCHOOT VAN DER GRACHT, W. A. J. M., JONGMANS, W. J. and TESCH, P., 1942, *Meded. Geol. Sticht.*, Serie C.1.2, No. 1, Maastricht.

VON KARMEN, T., 1911, Festgkeitsversuche unter allseitigem Druck, *Z. deutsch Ingenieure*, **245**, 433, 537.

VON RUSCH, H., 1959, *Zement-Kalk-Gips.* **12**, 1.

WELLMAN, H., 1950, Depth of burial of South Wales coals, *Geol. Mag.* **87**, 305.

WEST, W. D., 1939, The structure of the Shali "window" near Simla, *Geol. Surv. India Record.* **74**, 133.

WESTBROOK, J. H., 1958, Temperature dependence of strength and brittleness of some quartz structures, *J. Am. Ceram. Soc.* **41**, 433.

WILLIAMS, A., 1958, Oblique-slip faults and rotated stress systems, *Geol. Mag.* **95**, 207.

WILLIS, B., 1902, Stratigraphy and structure, Lewis and Livingstone ranges, Montana, *Geol. Soc. Am. Bull.* **13**, 305.

WOODWORTH, J. B., 1896, On the fracture systems of joints, with remarks on certain great fractures, *Boston Soc. Nat. History Proc.* **27**, 163.

INDEX